Avon

Collectible Fashion Jewelry and Awards

Monica Lynn Clements
and
Patricia Rosser Clements

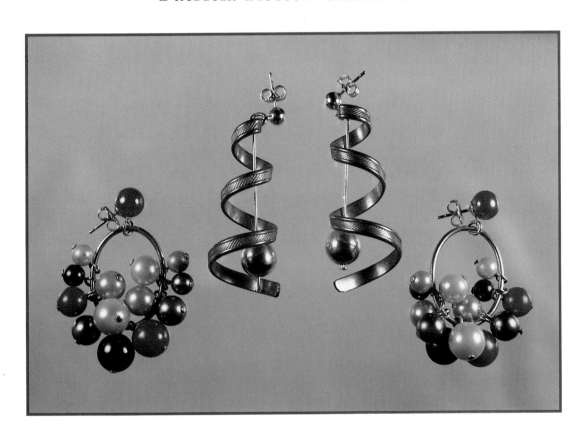

4880 Lower Valley Road, Atglen, PA 19310

Dedication

We dedicate this book to our friends and family
who have supported us in this endeavor.

Copyright © 1998 by Monica Lynn Clements and
 Patricia Rosser Clements
Library of Congress Catalog Card Number: 97-81265

All rights reserved. No part of this work may be reproduced or used in any form or by any means—graphic, electronic, or mechanical, including photocopying or information storage and retrieval systems—without written permission from the copyright holder.

Designed by Laurie A. Smucker
Typeset in Garamond/Embassy BT

ISBN: 0-7643-0523-9
Printed in China
1 2 3 4

Published by Schiffer Publishing Ltd.
4880 Lower Valley Road
Atglen, PA 19310
Phone: (610) 593-1777; Fax: (610) 593-2002
E-mail: schifferbk@aol.com
Please write for a free catalog.
This book may be purchased from the publisher.
Please include $3.95 for shipping.
Try your bookstore first.

We are interested in hearing from authors
with book ideas on related subjects.

Contents

Chapter One: History 7

Chapter Two: Representatives Awards and Gifts 27

Chapter Three: Sets 36

Chapter Four: Earrings 68

Chapter Five: Rings 83

Chapter Six: Pins 86

Chapter Seven: Bracelets 96

Chapter Eight: Necklaces 105

Chapter Nine: Related Awards and Gift 140

Endnotes 159

Bibliography 160

Acknowledgements

When we began this book, we had no idea we would encounter such an infinite variety of Avon jewelry designs. Our gratitude goes to friends, antique dealers, and Avon representatives who allowed us to photograph their collections.

We were fortunate to meet many Avon representatives who personified the spirit of Avon through their generosity. We thank all contributors whose names you will find in this book and those who wish to remain anonymous. A special thank you to Sarah Jones, who went that extra mile for us at what was a difficult and busy time for her. We are grateful to Miss Mary Jane, who allowed us to photograph her jewelry and her Mrs. Albee awards.

A special thanks goes to Mary G. Moon and Kenneth L. Surratt, Jr. for their diligence and enthusiasm.

Introduction

While the cost of collectible jewelry climbs, Avon fashion jewelry remains affordable. The variety and volume make it sought after by costume jewelry enthusiasts. More than ever before, Avon jewelry can be found at tag sales and in antique shops. The beauty, the versatility, and the affordability of Avon jewelry only ensure its popularity.

Detractors have snubbed the line because this fashion jewelry began with inexpensive pieces in the 1970s. The sheer volume of pieces distributed by Avon puts off some collectors. Yet changes have occurred that have stirred interest. While the company continues to provide affordable designs, it promotes styles that appeal to all tastes. Sterling silver pieces, Elizabeth Taylor Fashion Jewelry, and creations by designers of note such as Kenneth Jay Lane make Avon fashion jewelry more collectible than ever.

In the following pages, we present a wide selection of Avon fashion jewelry from the 1970s, 1980s, and 1990s. The lively motifs in the casual jewelry as well as the sophisticated designs showcase the wide appeal of this jewelry.

The final chapter of the book deals with the distinguished Mrs. Albee awards and related Avon awards. The Mrs. Albee figurines are awarded to members of the President's Club for sales achievement. President's Club members also receive the miniature Mrs. Albee awards. The Honor Society cups and saucers are examples of another line of awards that bear the likeness of Mrs. Albee.

The jewelry in the photographs is shown actual size unless otherwise noted. The purpose of this book is not to set firm prices but to serve as a guide. The prices reflect observations of the authors based on their experience in both buying and selling costume jewelry nationwide.

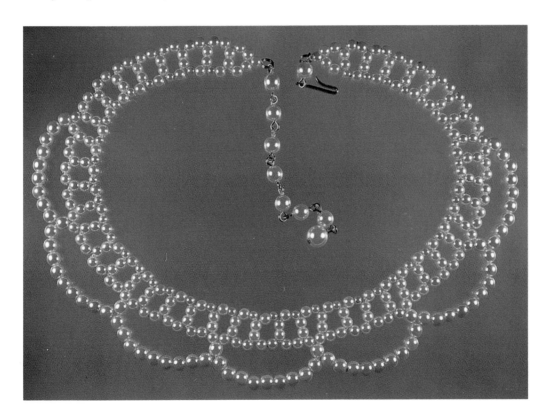

Chapter One
History

What began as the California Perfume Company (CPC) has become Avon Products, Inc., a name associated with quality goods and excellence in salesmanship. The company has expanded its line from makeup and toiletries to include home decor, clothing, and jewelry.

David Hall McConnell, who once worked for a publishing firm, formed CPC in the late 1800s. McConnell began his business by manufacturing five perfumes. He realized the importance of getting his products to the people who would buy them. His idea to market his perfume came from his experience in his former job when he sold books door-to-door.[1] McConnell's concept of direct marketing revolutionized the way women shopped and provided an independent means for CPC agents to make a living.

With the help of Mrs. Persis Foster Eames Albee of Winchester, New Hampshire, McConnell's enterprise became a major women's toiletries company. As the first "General Traveling Agent,"[2] Mrs. Albee possessed a gift for organizational skills. She recruited people to join the company as sales agents. Her diligence and success as the first door-to-door representative earned her the title, "Mother of the California Perfume Company."[3]

The name of Avon first appeared in 1928 with the introduction of talc, toothbrushes, a vanity set, and a cleaning product known as Avon Maid Cleanser.[4] The association of the Avon name continued with the company catalogs from the 1930s. In these publications, a comparison was made between the beauty of Shakespeare country in Stratford-upon-Avon in England and McConnell's home in Suffern, New York.[5] Promotional items with the name of Avon appeared in the following years. The popularity of the name caused it to become further linked with California Perfume Company when the company became known as CPC-Avon.

CPC-Avon changed and expanded with the times. By lowering prices on products and offering discounted specials, the company withstood The Depression.[6] The name of CPC-Avon became Avon Products, Inc. on October 6, 1939.[7] With the coming of World War II, Avon showed its versatility by producing such items as "insect repellent, pharmaceuticals, paratrooper kits, and gas mask canisters."[8] The company prospered after the war with its 65,000 representatives and $25 million in sales.[9] By the 1950s, Avon had established operations in Chicago, Kansas City, New York City, Pasadena, and Suffern. In 1954, a successful advertising campaign on television showcased the now familiar, "Ding-dong, Avon calling" slogan.[10]

The line of fashion jewelry officially began in 1971. Avon has marketed necklaces, bracelets, earrings, pendants, pins, and watches in a variety of designs at affordable prices. The creations have ranged from casual plastic jewelry to more formal-looking pieces. The growth of this line has made Avon a major distributor of costume jewelry.

The 1980s brought changes for Avon representatives. The workforce had diversified, and more women worked outside the home than ever before. The role of the Avon representative began to evolve. He or she no longer relied on going door-to-door as the sole method of selling products.[11]

As representatives worked to carve out a new niche for themselves, the company shifted its priorities. In an attempt to boost its status, Avon acquired Tiffany & Co. in 1984. Avon purchased Foster Medical, Mallinckrodt, the Mediplex Group, and Retirement Inns of America.[12] In a complete turn around, the company sold all the ventures and found itself the target of three takeover attempts.[13]

The unfortunate experiences of the 1980s changed Avon Products, Inc. for the better. The company has continued to develop alternative methods to sell products to the working woman through catalogs and the internet.[14] However, the Avon representatives make the difference in direct selling. They can be credited with over 95 percent of sales.[15] These workers have discovered resourceful ways to sell Avon products. For example, representatives in cities have relied on making sales at work or at community functions. In rural areas and in foreign countries, representatives tend to visit clients in their homes.[16]

Women continue to be a powerful force at all levels in the company. Staying competitive and working to sell goods in all parts of the globe, Avon Products, Inc. carries on the legacy that David McConnell began in 1886. Avon representatives make a difference as they make an independent living for themselves. Avon now boasts that the total number of sales representatives around the world is 2.3 million.[17] At the corporate level, women play an active role in shaping the future at Avon. Women hold 40 percent of the management jobs around the world.[18] Avon Products, Inc. remains stronger than ever as the company approaches the millenium.

The history of the Avon company represents a fascinating look at the changing role of women in business. Avon has donated its archives for study to the distinguished Hegler Museum and Library in Wilmington, Delaware. The museum plans to present the Avon collection for research in January of 1999.[19]

Representatives Awards and Gifts

McConnell worked to be an efficient motivator, and he wanted to bestow awards and prizes on individuals who excelled in his company. The founder of California Perfume Company saw the value of awards and incentives as a way to inspire loyalty and to reward the agents who persevered. The incentive jewelry that signified excellence was the Award Pin Program.[20] The first award pin was made of gold and inscribed with the initials, "CPC".[21]

The award system at Avon Products, Inc. has continued to evolve since 1910. McConnell's pin program has grown into the President's Club jewelry. Avon established the level of the President's Club in 1948 but did not award the first President's Club jewelry until 1973.[22] From this level, representatives have worked toward achieving higher sales goals and moving to Honor Society, Rose Circle, David H. McConnell, President's Council, and the President's Inner Circle.

The 4-A award debuted in 1961.[23] The four "A"s represent ability, achievement, ambition, and attitude.[24] Four curved Avon "A"s connected in the center along with a diamond set in the middle made up the design of this award pin for sales achievement. The 4-A pin was a reproduction of Avon's classic trademark symbol created in 1953 by Maxwell Rogers.[25] Later versions of this award included a 4-A pin with a sapphire in the center and similar pin with a pearl. The diamond and sapphire pins became the Golden Achievement Awards in 1971 and appeared again as part of the President's Club jewelry. A 4-A pin with a ruby in the center was the first award for representatives in 1973, who achieved the status of the President's Club.[26]

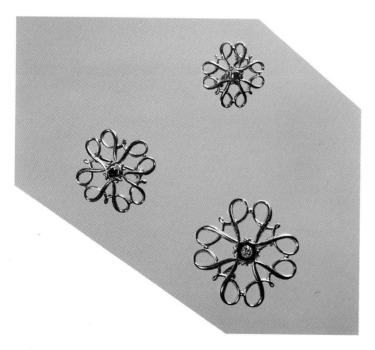

Lower center: Honor Society Pin, gold tone pin with sapphire, 1987. $35-45.
Bottom left: Avon Thousandaire Club lapel pin, 1986. $15-25.
Bottom center: Avon 100 Award Pin, gold tone, circa 1980s. $20-25.
Bottom right: Avon Thousandaire Club lapel pin, 1988. $15-25.
Award pins courtesy of Miss Mary Jane.

Opposite page:
Top: "A" Pin Award, circa 1970s. $15-20.
Top left: Gold Door Knocker Pin, circa 1960s. $25-30.
Top center: Avon 100 Pin, circa 1980s. $15-20.
Top right: Gold Door Knocker Pin, circa 1980s. $25-35.
Lower center: Honor Society Pin, gold tone pin with sapphire, circa 1987. $35-45.
Bottom left: Avon Thousandaire Club lapel pin, circa 1986. $15-25.
Bottom center: Avon 100 Award Pin, gold tone, circa 1980s. $20-25.
Bottom right: Avon Thousandaire Club lapel pin, circa 1988. $15-25.
Award pins courtesy of Miss Mary Jane.

Gold Rose Circle Award pin with rhinestones, circa 1980s. *Courtesy of Miss Mary Jane.* $35-45.

Top: Avon Identification pin, Balfour 10K MTG, circa 1945-1951. $35-40.
Left: Silver Avon Calling door knocker pin, 1964. $25-30.
Right: Gold Avon Calling door knocker pin, circa 1964-1983. $25-35.
Bottom: Queen's Award pin, 1964. $35-45.
Award jewelry courtesy of Kenneth L. Surratt, Jr.

Top left: David McConnell Award pin with rhinestone, 1990. $45-55.
Top center: Inner Circle Award Pin with rhinestones, circa 1990. $55-65.
Top right: David McConnell Award pin with porcelain likeness of Mrs. Albee, 1996. $45-55.
Bottom center: David McConnell Award in motif of Great Oak Tree with rhinestones, 1990. $65-75.
Award jewelry courtesy of Betty Barry.

Avon award jewelry comes in a variety of designs such as lapel pins, rings, pendants, pins, and timepieces. Through the design of selected award jewelry, Mrs. Albee has been immortalized. Her Victorian likeness adorns award jewelry, incentive items, and a line of awards that bear her name.

Other important symbols have found their way into award jewelry design. The acorn represents growth while the doorknocker is a symbol of trust. The Avon rose signifies beauty. The gold content or silver content in the pieces and the unique designs make the award jewelry valuable as well as collectible.

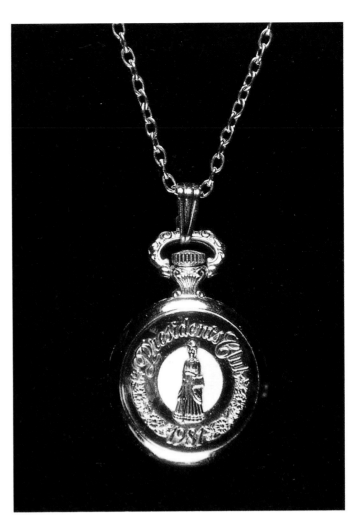

Back of watch with likeness of Mrs. Albee inscribed "President's Club 1981".

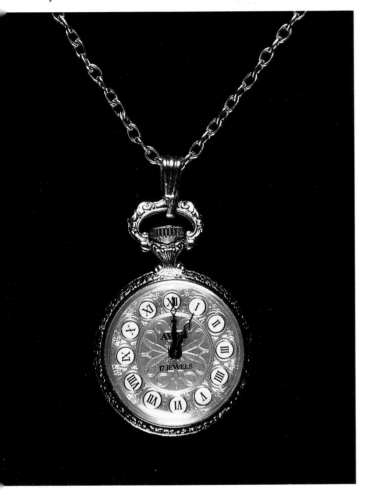

Avon President's Club gold watch on chain, chain measures 30 inches, 1981. Watch enlarged for photograph. *Courtesy of Martha A. McDougal.* $55-65.

Avon sterling silver award ring depicting Mrs. Albee, 1973. *Courtesy of Sarah Jones.* $25-35.

Enameled key chain depicting Mrs. Albee, 1987. *Courtesy of Virginia Young.* $20-30.

Avon Fashion Jewelry

Avon Products, Inc. distributes a vast amount of jewelry known as Avon fashion jewelry. Avon jewelry encompasses a wide range of styles. From informal plastic jewelry to the elegant and sophisticated look of designer pieces, Avon jewelry offers something for everyone.

A distinctive feature of Avon jewelry of the 1970s was glacé, a substance made of solid perfume. Pins, pendants, or rings opened to reveal the glacé center. This type of jewelry could also be found in pins designed for children. The glacé trend has continued through the 1990s with jewelry reminiscent of the earlier pieces.

An array of Avon fashion jewelry has a characteristic that exemplifies its versatility. A select number of earrings and necklaces have interchangeable or convertible pieces. Avon has designed parts, usually added or interchanged on earrings, that give the jewelry a different look and make it more flexible. On some pieces, Avon has given the wearer the option of converting an entire piece of jewelry. For example, a necklace can become a bracelet.

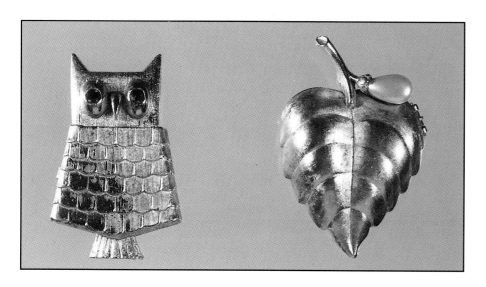

Left: Gold owl with emerald green stones for eyes glacé pin, circa 1970s. $20-25.
Right: Gold Leaf Pin Perfume Glacé pin with pearl, 1972. $20-25.
Pins courtesy of Mary G. Moon, Red Wagon Antiques.

Pins open to reveal glacé.

Top: Multi-colored plastic Avon bracelet, circa 1970s. $10-15.
Top right: Avon plastic earrings with interchangeable colored pieces, circa 1980s. $5-10.
Bottom: White plastic Avon bracelet, circa 1970s. $10-12.
Jewelry courtesy of Sarah Jones.

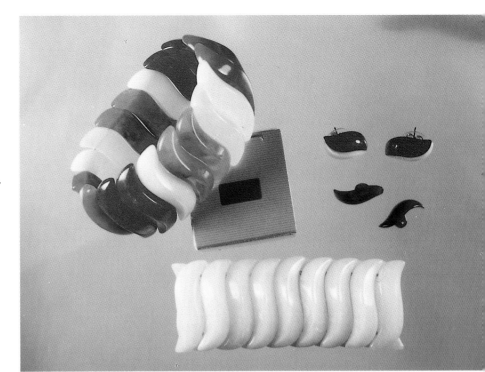

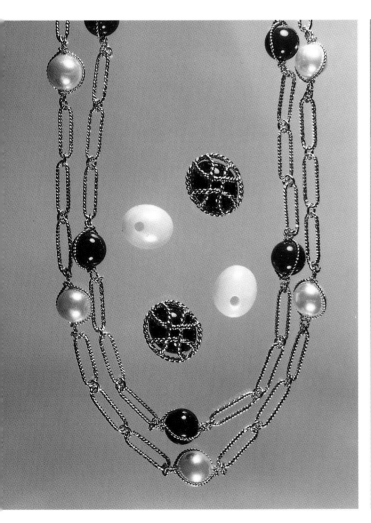

Necklace with pearls, 30 inches, circa 1990s. $15-20.
Necklace with lapis stones, 30 inches, circa 1990s. $15-20.
Earrings with lapis stones and interchangeable pearls, circa 1990s. $10-15.
Jewelry courtesy of Sarah Jones.

Plastic Sonnet Convertible Necklace/Bracelet, 24 inches, tassel 3 inches, 1973. *Courtesy of Mary Jo Michealis.* $25-35.

Popular motifs make Avon jewelry collectible. The seasonal pieces such as the Christmas jewelry bring customers back each year for new designs. Other holiday themes such as St. Patrick's Day, Valentine's Day, Mother's Day, and Father's Day have provided opportunities for Avon to offer special jewelry. Avon has delved into novelty jewelry with whimsical lapel pins in such designs as a hat, a basket, a crescent moon, or a cuckoo clock. Another popular subject includes children's jewelry.

Top: Enameled red and gold poinsettia pierced earrings with rhinestones, circa 1990s. $20-25.
Bottom center: Gold angel lapel pin with rhinestones in halo, circa 1990s. $15-20.
Bottom left and right: Sparkling Candle Earrings, circa 1990s. $10-15.
Jewelry courtesy of Miss Mary Jane.

Top left: Silver bell with bow lapel pin, circa 1980s. $15-20.
Top right: Enameled red, green, and black Candy Cane lapel pin, circa 1990s. $10-15.
Center left: Silver Christmas tree lapel pin, circa 1980s. $10-15.
Center: Enameled and gold poinsettia pin, circa 1980s. $15-20.
Center right: Silver and gold Partridge in a Pear Tree pin, circa 1980s. $12-15.
Bottom left: Gold Partridge in a Pear Tree lapel pin, circa 1990s. $20-25.
Bottom right: JOY lapel pin with rhinestones, circa 1990s. $20-25.
Pins courtesy of Miss Mary Jane.

Top: Gold arrow stickpin, circa 1980s. $15-20.
Center: St. Patrick's Day Hat's Off To Luck lapel pin with emerald green stones, circa 1980s. $10-15.
Bottom left: Gold lapel pin in shape of state of Arizona with "ARIZONA" engraved, circa 1990s. $20-25.
Bottom right: Gilded Bird lapel pin, 1976. $25-35.
Lapel pins courtesy of Kenneth L. Surratt, Jr.

Top left: Silver basket lapel pin, 1989. $15-20.
Top center: Cuckoo clock lapel pin, circa 1990s. $20-25.
Top right: Silver rabbit lapel pin with ruby red stone, 1989. $18-22.
Center left: Red and black enameled ladybug lapel pin, circa 1990s. $15-20.
Center: Plastic grapes lapel pin with gold leaves, circa 1990s. $15-20.
Center right: Moon lapel pin with rhinestone, circa 1980s. $15-20.
Bottom left: Gold teddy bear lapel pin, circa 1990s. $25-30.
Bottom center: Gold shoe lapel pin, circa 1990s. $15-18.
Bottom right: Plastic hat lapel pin with gold band, circa 1990s. $15-18.
Pins courtesy of Miss Mary Jane.

Ribbon and Hearts Mother's Day pin with ribbon and rose, 1990. $15-20.

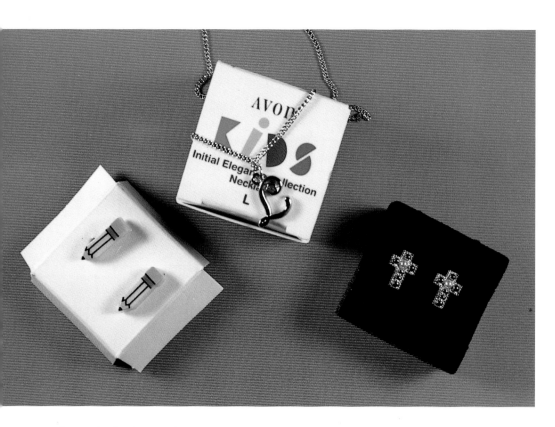

Left: Pencil School Days Clip-on Earrings, 1995. $5-10.
Center: Initial Elegance Collection "L" pendant with a ruby red stone, 1993. $6-12.
Right: Avon Precious Cross Pierced Earrings with seed pearls and gold balls, 1995. $5-10.
Children's jewelry courtesy of Miss Mary Jane.

Ballerina Charm Bracelet, circa 1990s. $10-15.

Patriotic jewelry comes in a myriad of designs from commemorative pins to fashion jewelry. Lapel pins commemorating the Centennial Liberty Campaign in 1986 showed Avon's commitment to donating money for the refurbishment of the Statue of Liberty.[27] As a sponsor for the summer Olympics, the company produced colorful commemorative jewelry. Some designs depict a patriotic motif with a variation of stars and/or stripes.

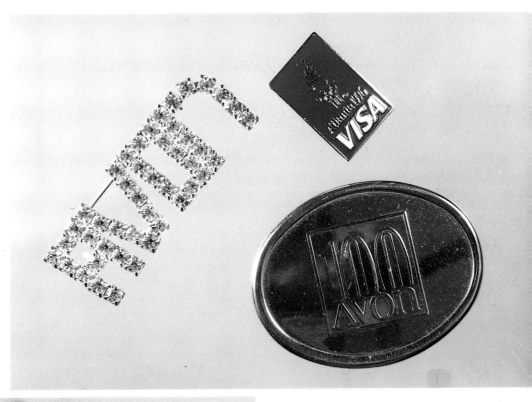

Left: Avon Sales Achievement gift pin made of rhinestones, circa 1980s. $25-35.
Top: VISA Olympic Award lapel pin commemorating Avon's sponsorship of the Summer Olympics, 1996. $10-15.
Bottom: Avon 100 Award, small silver box, 1986. $20-30.
Jewelry courtesy of Sarah Jones.

Left: Enameled pin commemorating Statue of Liberty, 1886-1986. $15-20.
Bottom center: USA enameled pin, 1986. $10-15.
Right: Avon Centennial anniversary mirror, 1986. $10-15.
Items courtesy of Virginia Young.

Three pairs of Sparkle Star Pierced Earrings with rhinestones, ruby red stones, or sapphire colored stones, circa 1990s.
Courtesy of Sarah Jones. $15-25.

Figurals are popular with collectors. Avon Products, Inc. presents an array of animals from cats, teddy bears, kangaroos, and koala bears to mice. In addition, frogs, birds, and butterflies make lively subjects for jewelry. Figurals come in different sizes and can be found as pins, lapel pins, necklaces, or sets.

Left: Puppy Love Pin Pal Glacé, 1975. $10-20.
Center: Christmas puppy pin, circa 1970s. $10-15.
Right: Willie the Worm Pin Pal Glacé, 1974. $15-20.

Left: Mouse pin, circa 1990s. *Courtesy of Doris Vaughan.* $15-20.
Right: Mouse pin, circa 1990s. $15-20.
Bottom left: Owl pin, 1975. $15-20.
Bottom right: Kangaroo pin, circa 1980s. $15-20.

Left: Frisky Kitty Pin with bow, bell, and sapphire blue eyes, 1975. *Courtesy of Merle Fellows.* $20-25.
Center: Avon Cat Eyeglass Holder Pin with rhinestone eyes, 1996. $10-15.
Right: Gold pin with cat motif, circa 1990s. $10-15.
Unless otherwise noted, jewelry in this photograph is courtesy of Sarah Jones.

Assorted frog lapel pins, small gold frog in middle is an earring, lapel pins are circa 1980s-1990s, earring is circa 1995. Lapel pins are $15-20, pair of earrings is $8-10.

Top left: Green enameled dragonfly lapel pin, circa 1980s. $15-18.
Top right: Small gold bird lapel pin, circa 1980s. $15-18.
Center: Gold hummingbird pin with ruby red stone, circa 1990s. $25-35.
Bottom left: Butterfly lapel pin, circa 1980s. $12-16.
Bottom right: Gilded bird lapel pin, 1976. $25-35.
Pins courtesy of Miss Mary Jane.

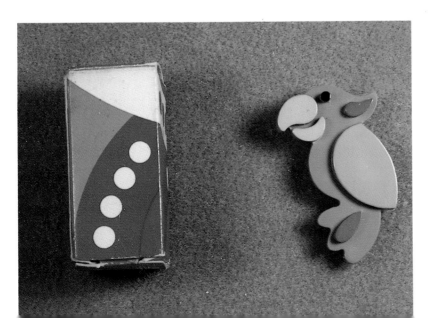

Perky Parrot pin and box, can be worn as pendant, 1973. *Courtesy of Mary G. Moon.* $15-20.

While the chains or the pendant necklaces are traditional accessory choices, Avon offers other alternatives. The lariat necklace provides versatility by allowing the wearer to choose the length and shape of the necklace. Slide necklaces with different motifs have wide appeal. Enhancers attached to the necklace change the look of the jewelry. Another option in selected pieces of Avon jewelry is the possibility of joining two necklaces together to form one long necklace.

Left: Loop over lariat necklace with jade stones, 15 inches, circa 1980s. $30-40.
Right: Enameled lapis pendant with floral motif, 16-inch chain, circa 1980s. $35-45.
Necklaces courtesy of Miss Mary Jane.

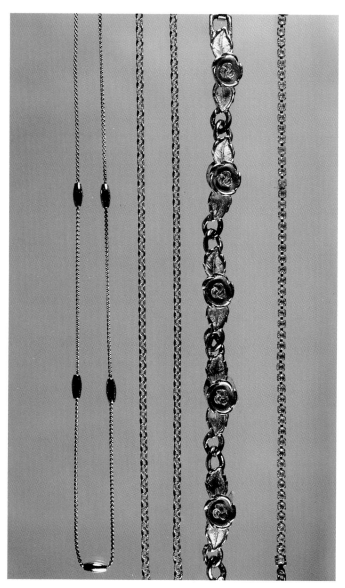

Left: Silver necklace with lapis beads, circa 1980s. $20-30.
Center left: Gold chain, circa 1980s. $18-22.
Center right: Bracelet with enameled flowers and gold leaves, circa 1980s. $25-35.
Right: Gold chain, circa 1980s. $15-20.
Necklaces and bracelet courtesy of Kenneth L. Surratt, Jr.

Left: Silver cross with aquamarine stone, 20-inch chain, circa 1990s. $25-35.
Center: Lariat necklace with two silver bells, 24-inch chain, circa 1980s. $20-30.
Right: Gold pendant with cabochons, 20-inch chain, circa 1990s. $20-25.

The designers at Avon Products, Inc. have innovative ideas about how to accessorize. They provide jewelry that goes with sets or pieces from previous years. Popular designs and variations on popular designs return year after year. These combinations with additional creations keep the look of the jewelry fresh from season to season.

Left: Gold slide necklace with bee and pearls, 20-inch chain, circa 1980s. $25-35.
Center: Gold pendant with four hearts, 18-inch chain, circa 1990s. $25-35.
Right: Gold slide necklace with turtle, chain measures 26 inches, circa 1990s. $25-35.

Necklace and earrings with multi-colored beads, 30-inch necklace, circa 1990s. $15-25.
Top: Bracelet with colored stones, circa 1990s. $10-15.
Bottom: Brooch with colored stones, circa 1970s. $15-20.
Jewelry courtesy of Sarah Jones.

Top left: Enhancer with hematite stone, circa 1980s. $25-35.
Top right: Gold Imari enhancer with clasp, 1985. $20-25.
Bottom: Gold Renaissance Beauty Clasp enhancer, 1987. $20-25.
Enhancers courtesy of Sarah Jones.

Cat Bracelet and Two Necklaces, 20-inch outer necklace, 17-inch inner necklace, can be hooked together to form one long necklace, circa 1980s. *Courtesy of Sarah Jones.* $40-50.

Tortoiseshell 30-inch necklace, earrings, and piano pin. Necklace and earrings, circa 1970s, and pin, circa 1990s. *Courtesy of Sarah Jones.* Necklace and earrings are $25-35; piano pin is $15-20.

Avon has utilized jewelry to support the cause of Breast Cancer Awareness. The Pink Ribbon pieces consist of ribbon-shaped enameled and gold lapel pins in two sizes, earrings, and a writing pen. The proceeds from these items go to breast cancer education in communities throughout the United States.[28]

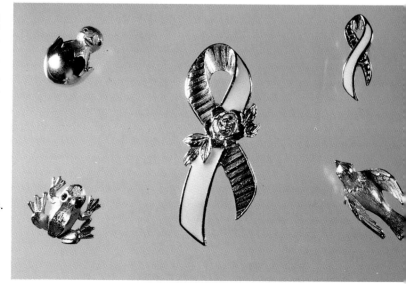

Top left: Chicken and egg lapel pin, circa 1970s. $15-20.
Top right: Pink Ribbon Breast Cancer Awareness lapel pin, circa 1990s. $5-10.
Bottom left: Frog lapel pin, circa 1990s. $10-15.
Center: Pink Ribbon Breast Cancer Awareness lapel pin, circa 1990s. $5-8.
Bottom right: Gilded bird lapel pin, 1976. $25-35.
Jewelry courtesy of Doris Vaughan.

Avon jewelry designs appear not only for women but also for men. Lapel pins, tie tacks, bracelets, and watches are examples of the items available for men. Avon offers pendant necklaces designed exclusively for men. The designs come in novelty jewelry and figurals. The arrowhead jewelry is an example of the Western motif.

Three men's tie tacks, 1985. $15-20 each.

Left: Men's necklace with arrow and Indian head coin, gold 28-inch chain, circa 1980s. $45-55.
Center: Men's necklace with polished stone arrowhead, gold 30-inch chain, circa 1980s. $40-50.
Right: Wooden arrow, gold 28-inch chain, circa 1980s. $20-25.
Necklaces courtesy of Martha A. McDougal.

Left: Buffalo pewter men's necklace, silver 30-inch chain, circa 1980s. $25-35.
Center: Men's necklace with scrimshaw pendant depicting Bengal tiger in ink on glass, 32-inch chain, circa 1980s. $35-45.
Right: Men's gold necklace with tooth pendant, 30-inch chain, circa 1980s. $20-30.
Necklaces courtesy of Martha A. McDougal.

Right: Man's water resistant quartz watch, circa 1990s. $55-65.
Far Right: Man's water resistant quartz watch, circa 1990s. $45-55.
Watches *courtesy of Sarah Jones.*

Jewelry by noted designers has brought something special to Avon fashion jewelry. Kenneth Jay Lane's designs in sets and single pieces have created an element of sophistication. Elizabeth Taylor chose Avon to showcase her elegant line of jewelry. In 1993, this line included costume jewelry modeled after Taylor's own jewelry and pieces she wore in noted films. The Smithsonian Institution Series, with designs based upon art treasures at the museum, has made classic jewelry available.

Kenneth Jay Lane Nature's Treasures pin with enameled wings, pink stone, and rhinestones, 1990. *Courtesy of Miss Mary Jane.* $35-45.

Taylored Style Collection Small Necklace and Clip-on Earrings, 16-inch necklace, 1995. *Courtesy of Sarah Jones.* $95-125.

Queen Elizabeth I Pierced Earrings from the Smithsonian Series with ruby stones and pearls, 1996. *Courtesy of Sarah Jones.* $20-30.

Left: Elizabeth Taylor Evening Star Watch with rhinestones and pearls, circa 1995. $95-125.
Right: Two Elizabeth Taylor Brilliance 22K Gold Rings with rhinestones, circa 1996. $75-95 each.
Jewelry courtesy of Sarah Jones.

Smithsonian Angel Pin with green, red, and gold enamel, 1996. *Courtesy of Sarah Jones.* $20-30.

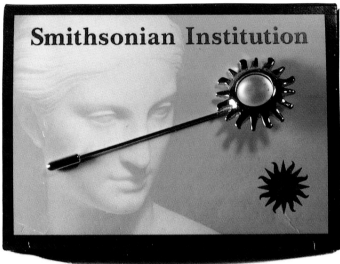

Sun Emblem Sunburst Stickpin with gold tone and frosted cabochon from the Smithsonian Institution Series, 1996. *Courtesy of Miss Mary Jane.* $25-35.

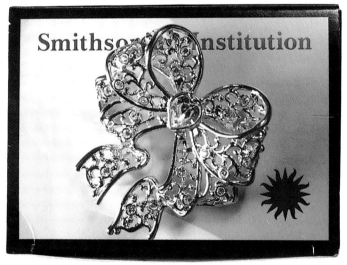

Open Work Bow Pin from the Smithsonian Series with gold and rhinestone, circa 1990s. *Courtesy of Sarah Jones.* $25-35.

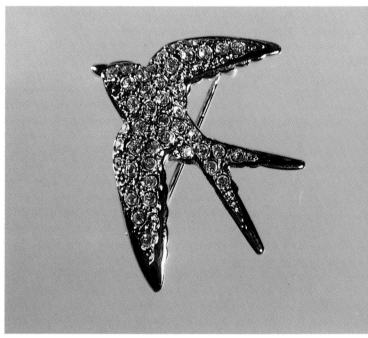

Small Juliette Gordon Low Swallow of rhinestones and silver and sapphire blue eye from the Smithsonian Institution Series, circa 1996. Enlarged for photograph. *Courtesy of Miss Mary Jane.* $25-35.

While the jewelry of the 1970s was inexpensive, Avon has since expanded its jewelry line. In the 1980s and 1990s, new designs and designer pieces were added. Avon has continued its tradition of offering fun jewelry as well as choices that are more formal. Avon continues to be one of the largest distributors of costume jewelry in the world. While the jewelry is wide ranging in styles and motifs, the volume has not compromised the quality of the designs.

Chapter Two
Representatives Awards and Gifts

Smile Pendant Award, red enamel, gold 14-inch chain, 1978. *Courtesy of Mary G. Moon.* $20-25.

Dedicated to Service 1886-1962 Award bracelet. *Courtesy of Kenneth L. Surratt, Jr.* $25-35.

Centennial Award necklace with rhinestones, signed "1886 AVON 1986", 15 inches. *Courtesy of Mary G. Moon.* $25-35.

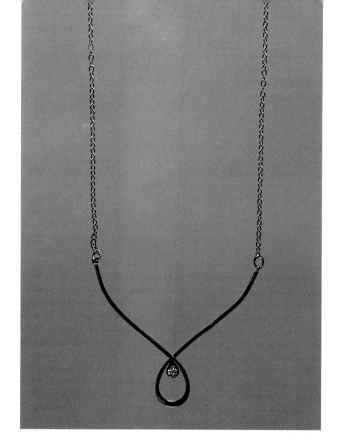

Avon Managers Gift Diamond Loop necklace, 14 karat gold, 1977. $45-55.

Avon Sales Achievement rhinestone gift pin, circa 1980s. *Courtesy of Virginia Young.* $25-35.

Blue and White Honor Award plastic cameo locket/pendant necklace depicting Mrs. Albee, 1966.

Moonwind Award glass intaglio depicting Artemis (Diana), goddess of the hunt, with twisted silver frame, 20-inch chain, 1975. *Courtesy of Mary G. Moon.* $25-35.

Inside of Honor Award cameo with inscription, "Honor Award President's Campaign 1966."

Top: Honor Society Award pin with rhinestones, circa 1990s. $25-35.
Bottom: Honor Society Award pin with aquamarine blue stone, circa 1990s. $20-30.
Right: Honor Society Award, 1990. $35-45.
Achievement awards courtesy of Sarah Jones.

Top: Avon gold sales achievement gift pin with rhinestones, circa 1990s. $25-35.
Center top. Honor Society Award pin with ruby stone, 1987. $25-35.
Center: President's Club Achievement Award brass pin with three pearls, 1985. $35-45.
Bottom: Honor Society Award pin with ruby stone, 1988. $45-55.
Achievement awards courtesy of Sarah Jones.

Top: President's Achievement Award brass pin with three sapphire blue stones, 1986. $35-45.
Center: President's Club Star Pin Award with red star, 1989. $30-35.
Bottom: President's Club Star Pin Award with pearl star and blue and white star, 1989. $45-55.
Awards courtesy of Sarah Jones.

Top: Lapel pin commemorating Parfums Paris, circa 1990s. $15-20.
Top center: Avon $1,000 Club lapel pin, 1985. $20-25.
Right: Gold tone President's Club Pin award, 1982. $20-25.
Bottom left: Rising Star tie tack, 1990. $15-20.
Bottom center: Honor Society award pendant with porcelain portrait of Mrs. Albee, 1990-1997. $45-55.
Jewelry courtesy of Sarah Jones

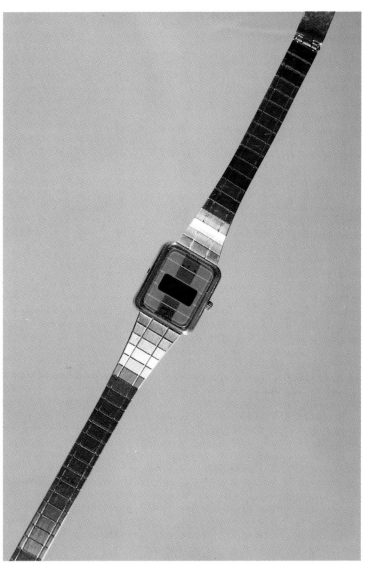

Winning is Beautiful Avon award watch, circa 1980s. *Courtesy of Sarah Jones.* $35-45.

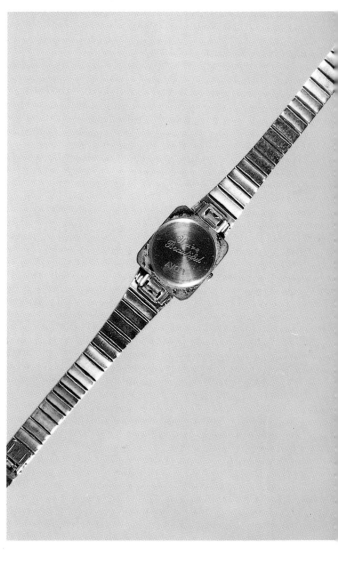

Back of Avon award watch inscribed, "You're Beautiful".

Honor Society Award Clock, battery operated quartz clock with congratulations message, 1987. *Courtesy of Sarah Jones.* $25-35.

Avon Centennial jewelry set, includes gold necklace, bracelet, and earrings with rhinestones, 1986. *Courtesy of Sarah Jones.* $175-195.

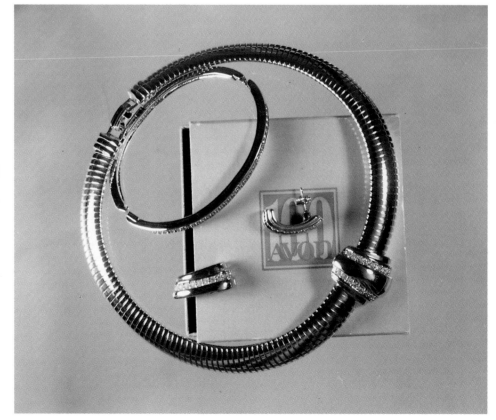

Top left: David McConnell Award pin with porcelain likeness of Mrs. Albee, 1991. $45-55.
Top center: Blue and silver Rising Star Award lapel pin, circa 1980s. $15-20.
Top right: David McConnell Great Oak Tree Award with rhinestones, 1990. $65-75.
Center: David McConnell Award with rhinestone, 1990. $45-55.

Bottom left: Inner Circle Award pin with rhinestones, 1990. $55-65.
Bottom center: Sponsorship Award 10K bracelet with sapphire stone, 1986. $50-60.
Bottom right: Enameled lipstick lapel pin commemorating Avon's sponsorship of the 1996 Summer Olympics, 1996. $15-25.
Award jewelry courtesy of Betty Barry.

Enlarged image of Inner Circle Award pin, circa 1990.

Opposite page:
Top left and right: Rising Star award lapel pins, circa 1980s. $15-20 each.
Top left: The President's Celebration award pendant, 1980. $20-30.
Top center: President's Club Pin Award, gold tone pin with "80" in center surrounded by 4-A design, 1980. $20-25.
Top center: President's Club Pin Award, gold tone pin with "79" in center surrounded by 4-A design, 1979. $20-25.
Top right: President's Club Pin Award, gold tone pin with "81" in center surrounded by 4-A design, 1981. $20-25.
Center left: President's Club Star Pin Award with pearl star and blue and white star, 1989. $45-55.
Center: Honor Society Sales Achievement Award, 1990. $35-45.
Center right: President's Club Star Pin Award with gold star, pearl star, and white star, 1990. $75-85.
Bottom right: Honor Society 100 Pin, gold tone, circa 1980s. $25-35.
Bottom left: President's Club Award gold tone stickpin, 1981. $20-25.
Bottom center: Rose Circle Award pin depicting Mrs. Albee, 1990s. $45-55.
Bottom right: Rose Award, 1985. $15-25.
Award jewelry courtesy of Miss Mary Jane.

President's Club Award timepiece depicting Mrs. Albee, 1981. *Courtesy of Miss Mary Jane.* $55-65.

Back of President's Club Award timepiece with likeness of Mrs. Albee inscribed, "President's Club 1981".

Top right: Sponsorship Award gold pin, 1986. $30-40.
Bottom: Sponsorship Award 10K bracelet with sapphire stone, 1986. $50-60.
Award jewelry courtesy of Betty Barry.

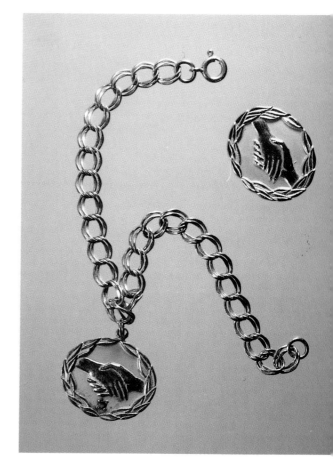

Top left: Gold Rose Circle Award pin with rhinestones, circa 1990s. $30-35.
Top center: Honor Society Award gold pin with sapphire, 1988. $25-35.
Top right: Gold Honor Society Award pin with rhinestones, circa 1990s. $35-45.
Bottom left: President's Club Achievement Pin Award, brass pin with four sapphire blue stones, 1986. $45-55.
Bottom center: President's Club Achievement Pin Award, brass pin with four ruby red stones, 1984. $45-55.
Bottom right: President's Club Achievement Pin Award, brass pin with four pearls, 1985. $45-55.
Award pins courtesy of Miss Mary Jane.

Top: Avon Sales Representative Gift pin, gold with rhinestones in heart, circa 1980s. $25-35.
Center: Avon Sales Representative Gift pin, gold with rhinestones, circa 1990s. $30-40.
Bottom: Avon Sales Representative Gift rhinestone pin, circa 1980s. $25-35.
Gift pins courtesy of Miss Mary Jane.

Gold coin 90th Anniversary Bicentennial Pendant Award, 28-inch chain, 1976. *Courtesy of Miss Mary Jane.* $65-95.

Rose Circle Award Pin with ruby red stone, 1-inch square, circa 1990s. Enlarged for photograph. *Courtesy of Miss Mary Jane.* $25-35.

Back of 90th Anniversary Bicentennial Pendant Award.

Chapter Three
Sets

Mother of pearl necklace and earrings, Mother's Day 1986. *Courtesy of Virginia Young.* $20-30.

Necklace and earrings, black and white plastic with gold, circa 1980s. *Courtesy of Virginia Young.* $20-25.

Pearl and gold set with earrings, brooch, and bracelet, 1971. $35-45.

Avon earrings and necklace from Blue Box Collection, 34-inch chain, .38-inch earrings, circa 1980s. *Courtesy of Kenneth L. Surratt, Jr.* $20-25.

Necklace and earrings with pink tulips, 20-inch gold-tone chain, 1.75-inch earrings, 1994. *Courtesy of Kenneth L. Surratt, Jr.* $15-25.

Top: Avon earrings convertible to pierced earrings with natural agate stones, circa 1990s. $10-15.
Bottom: Desert Stones necklace with natural agate and carnelian stones, 19-inch chain with gold balls and stones, circa 1990s. $20-25.
Earrings and necklace courtesy of Kenneth L. Surratt, Jr.

Set with earrings and necklace of gold metal with emerald stones and pearls, 23-inch chain, circa 1980s. $25-35.

Left and right: Sterling silver cubic zirconia earrings, 1997. $20-25.
Bottom: Sterling necklace with cubic zirconia, 1997. $20-30.
Earrings and necklace courtesy of Kenneth L. Surratt, Jr.

Avon set with rhinestone pierced earrings and "M" pendant, 10-inch chain, circa 1980s. $15-20.

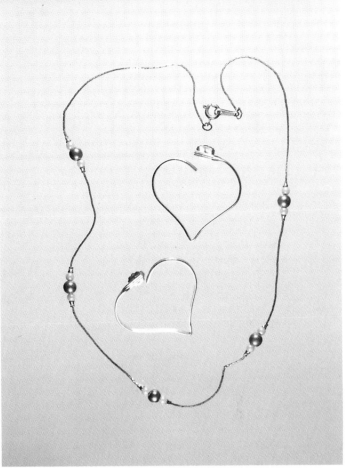

Gold 16-inch chain with groups of pearls, circa 1990s. $15-20.
Gold heart-shaped pierced earrings, circa 1990s. $10-15.

Abolone set includes necklace with 14-inch silver metal chain, silver bracelet trimmed with gold, and ring set in silver, circa 1990s. $35-45.

Necklace of plastic white and gold, 18 inches, circa 1980s. $15-20.
Top center: Pink Lucite with pearl pierced earrings, circa 1990s. $10-15.
Bottom center: Plastic white pierced earrings, circa 1980s. $10-12.
Jewelry courtesy of Kenneth L. Surratt, Jr.

Gold and plastic necklace and bracelet set, circa 1980s. $20-25.

Top: Rose petal porcelain pin, circa 1987. $20-25.
Bottom: Two pair of porcelain petal earrings, 1987. Each pair $10-15.
Jewelry courtesy of Virginia Young.

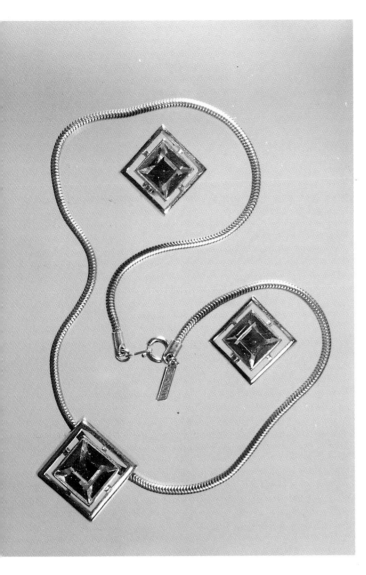

Faceted stone necklace and earrings set, 15-inch chain, 1977. *Courtesy of Lila Heather.* $25-35.

Necklace and earrings set with jade cabochons and gold, circa 1970s. $25-35.

Avon Pasted Blossoms set of white plastic with pearls necklace, and pierced earrings, 15-inch necklace, 1986. *Courtesy of Virginia Young.* $20-25.

Summer berries set with necklace and earrings with iridescent sheen, 16-inch necklace, circa 1990s. *Courtesy of Sarah Jones.* $20-35.

Plastic earrings and elasticized bracelet in pastel colors, circa 1980s. *Courtesy of Sarah Jones.* $15-20.

Opaque necklace and pierced earrings set, 20-inch necklace, circa 1990s. *Courtesy of Sarah Jones.* $20-25.

Plastic multi-colored bracelet and pierced earrings with floral designs, circa 1990s. *Courtesy of Sarah Jones.* $15-20.

Plastic necklace and earrings set, 18-inch necklace, circa 1990s. *Courtesy of Sarah Jones.* $20-25.

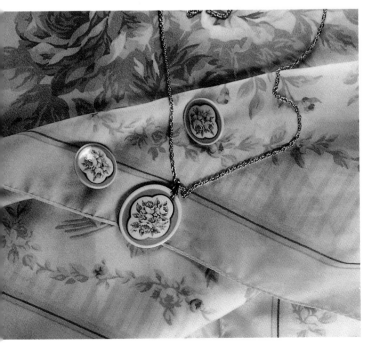

Avon floral scarf and Avon earrings and necklace, 24-inch chain and pendant measures 1.38 x 1.12 inches, scarf is 34 inches, circa 1990s. *Courtesy of Sarah Jones.* Scarf is $25-35; earrings and necklace are $15-18.

Avon plastic necklace, bracelet, and earrings set, 10-inch necklace, circa 1990s. *Courtesy of Sarah Jones*. $20-25.

Avon scarf with belt and two pairs of earrings, scarf is 38 x 38 inches, circa 1980s. *Courtesy of Sarah Jones*. Scarf, $25-35; belt and matching earrings, $15-25; plastic earrings, $10-15.

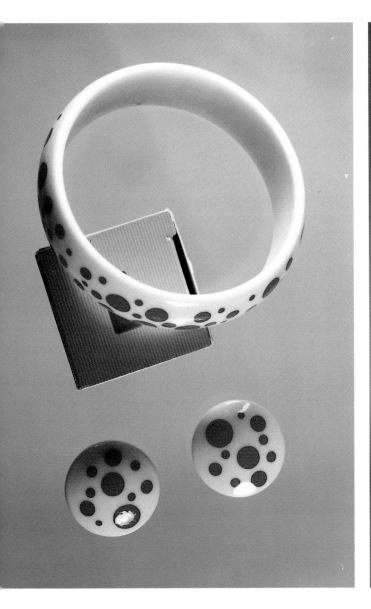

White and red plastic bracelet and earrings, circa 1980s. *Courtesy of Sarah Jones.* $15-20.

Plastic necklace and earrings, 18-inch necklace, circa 1990s. *Courtesy of Sarah Jones.* $20-25.

Red and white Avon purse and plastic set with bracelet, pierced earrings, and 30-inch necklace, purse circa 1980s, jewelry set, circa 1990s. *Courtesy of Sarah Jones.* Purse is $15-25; jewelry set is $25-35.

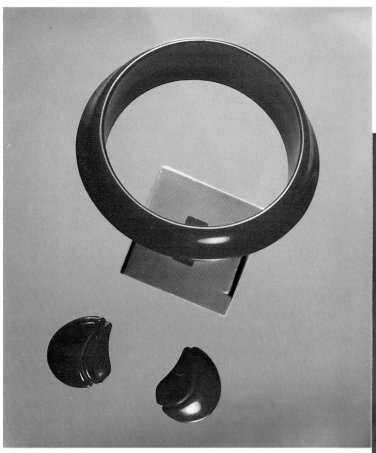

Red plastic pierced earrings and bracelet, circa 1990s. *Courtesy of Sarah Jones.* $15-20.

Amber colored plastic earrings and 24-inch necklace with amber and white beads, circa 1980s. *Courtesy of Sarah Jones.* $20-25.

Avon 34-inch scarf with pearl jewelry design, circa 1990s. $35-45. *Center and right:* Summit Collection Earrings and Necklace with dangle pearls, 18-inch necklace, circa 1990s. $35-45. *Scarf and jewelry courtesy of Sarah Jones.*

Sienna-colored plastic and metal 26-inch necklace and earrings, circa 1990s. *Courtesy of Sarah Jones.* $15-25.

Multi-colored plastic necklace and two pairs of plastic earrings, 19-inch necklace, circa 1970s. *Courtesy of Sarah Jones.* $25-35.

Avon 34-inch scarf with floral motif, circa 1980s. $45-55
Plastic necklace with metallic spheres and amber, white, and red colored beads and matching plastic and metal pierced earrings, 25-inch necklace, circa 1980s. $25-35.
Scarf and jewelry courtesy of Sarah Jones.

Two plastic yellow and white necklaces and one pair earrings, necklaces measure 16 and 20 inches, circa 1980s. *Courtesy of Sarah Jones*. Necklaces, $15-20 each; earrings, $10-15.

Plastic earrings and 20-inch necklace with seashell design, circa 1990s. *Courtesy of Sarah Jones*. $15-18.

Avon silk scarf with leopard design, circa 1990s. $50-60.
Top right: Gold earrings with animal design, circa 1990s. $20-25.
Top left: Metal pierced leopard earrings, circa 1970s. $15-20.
Bottom left: Gold giraffe pierced earrings, circa 1990s. $15-20.

Plastic green and white necklace and earrings, 30-inch necklace, circa 1990s. *Courtesy of Sarah Jones.* $20-25.

Black earrings with rhinestones and black tie pin with rhinestones, circa 1986. *Courtesy of Sarah Jones.* $15-20.

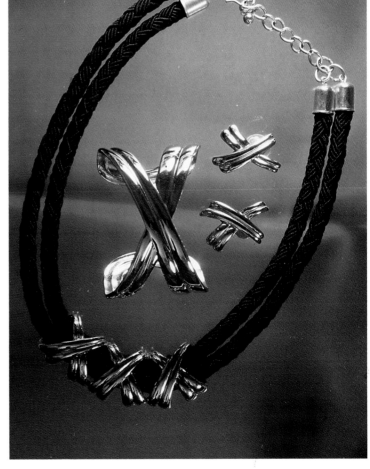

Bracelet, 18-inch necklace, and earrings set, circa 1990s. *Courtesy of Sarah Jones.* $25-30.

Colorful 24-inch necklace and black earrings, circa 1990s. *Courtesy of Sarah Jones.* $20-25.

Necklace with lapis beads and earrings with lapis, 24-inch necklace, circa 1990s. *Courtesy of Sarah Jones.* $25-30.

Black and white Avon 34-inch scarf, circa 1980s. $25-35.
Black and white plastic bracelet and pierced earrings, circa 1980s. $15-20.
Scarf and jewelry courtesy of Sarah Jones.

Necklace with green plastic beads, pearls, and green plastic earrings, 16-inch necklace, circa 1980s. *Courtesy of Sarah Jones.* $15-20.

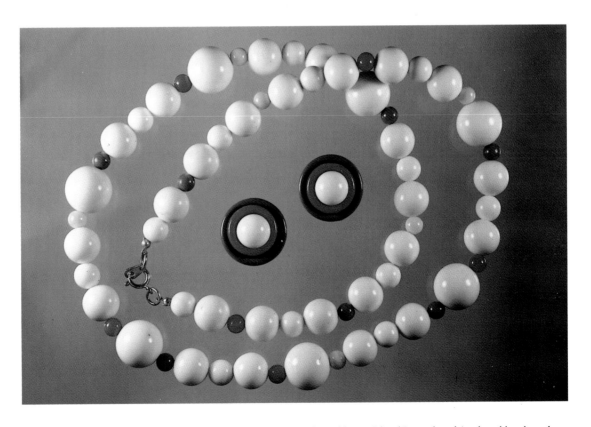

Plastic 30-inch necklace with white and multi-colored beads and earrings with interchangeable pieces and removable rings, circa 1980s. *Courtesy of Sarah Jones.* $25-30.

Necklace, bracelet, and earrings of off-white plastic, 16-inch necklace, circa 1980s. *Courtesy of Sarah Jones.* $20-25.

Black plastic bead necklace with enhancer and earrings with rhinestones, necklace (doubled for photograph) is 60 inches, circa 1980s. *Courtesy of Sarah Jones.* $25-35.

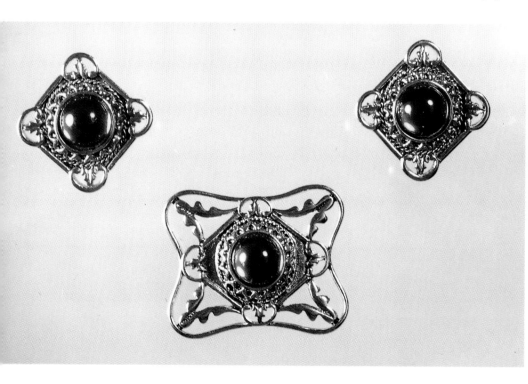

Bold and Classic gold metal earrings and brooch with amber-colored stones, circa 1990s. *Courtesy of Sarah Jones.* $20-25.

Sand and Sea Shell Bracelet and gold earrings with pearls and shell motif, circa 1990s. *Courtesy of Sarah Jones.* $20-25.

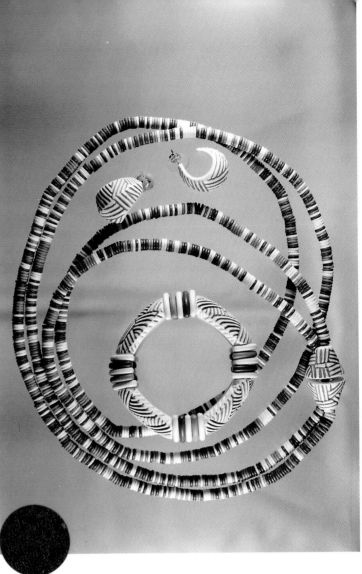

Shell necklace, elasticized bracelet, and earrings, necklace (coiled to show clasp) is 30 inches, circa 1990s. *Courtesy of Sarah Jones.* $18-22.

Enameled earrings and brooch with floral motif of pansies, circa 1980s. *Courtesy of Sarah Jones.* $15-25.

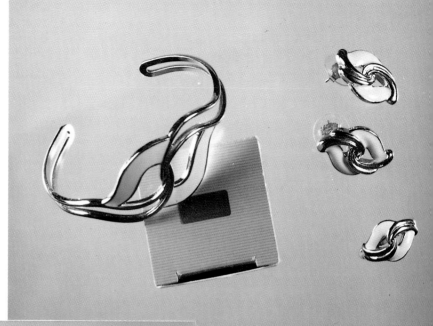

Key Biscayne bracelet, earrings, and ring set, circa 1980s. *Courtesy of Sarah Jones.* $22-28.

Green enameled bracelet and pierced earrings, circa 1990s. *Courtesy of Sarah Jones.* $18-22.

Left: Gold bracelet with gold charms depicting seashells, starfish, sea horse, and sand dollar, circa 1990s. $20-25.
Right: Gold star pierced earrings with pearls, gold shells, and sea horse, circa 1990s. $15-20.
Jewelry courtesy of Sarah Jones.

Yellow metal bracelet and pierced earrings, circa 1980s. *Courtesy of Sarah Jones.* $15-18.

Necklace with gold chain, pearls, and leaves, 30-inches, circa 1990s. $20-25.
Top: Gold squirrel lapel pin with pearl, circa 1990s. $15-20.
Bottom: Gold earrings depicting leaves and acorns, circa 1990s. $10-15.
Jewelry courtesy of Sarah Jones.

Top: Plastic cameo pierced earrings, circa 1994. $10-15.
Bottom: Avon Romantic Poet Necklace with plastic cameo pendant and pink stones, 30-inch chain, circa 1990s. $25-35.
Jewelry courtesy of Sarah Jones.

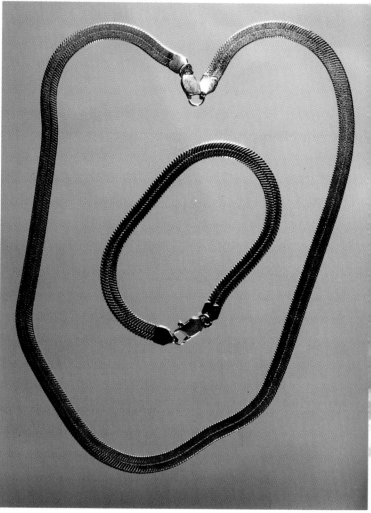

Gold 18-inch necklace and bracelet, circa 1980s. *Courtesy of Sarah Jones.* $25-35.

Onyx and rhinestones set with heart pendant, ring, and earrings, 18-inch chain, circa 1990s. *Courtesy of Sarah Jones.* $45-55.

Tri-colored 16-inch necklace and bracelet, circa 1980s. *Courtesy of Sarah Jones.* $25-35.

Avon Heirloom Rose Bracelet and Earrings, circa 1980s. *Courtesy of Sarah Jones.* $20-25.

Rose Passion Pin and Earrings with rhinestones, circa 1990s. *Courtesy of Sarah Jones.* $35-45.

Gold and red enameled "NOEL" necklace and earrings, 26-inch chain, circa 1990s. *Courtesy of Kim Griffin.* $25-35.

Sparkling Accents Hoop Earrings and Ring with rhinestones, circa 1980s. *Courtesy of Sarah Jones.* $25-35.

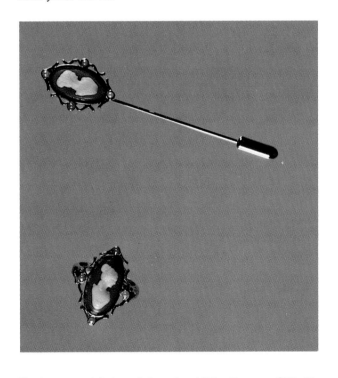

Plastic cameo stickpin and ring, circa 1980s. *Courtesy of Miss Mary Jane.* $25-35.

Set with necklace and earrings of gold with pearls and rhinestone, circa 1990s. $20-30.

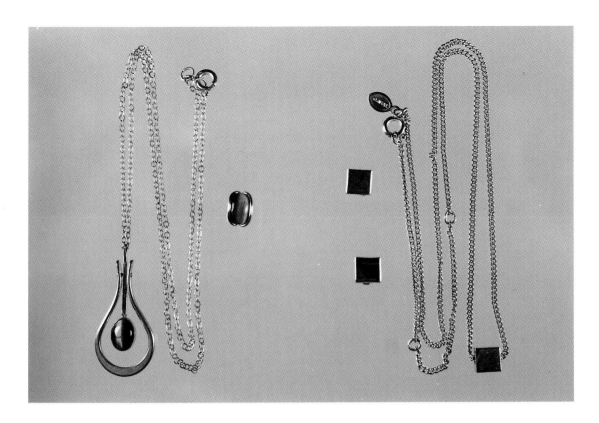

Left: Necklace and earring with tiger eye stone, 19-inch chain, circa 1980s. $20-30.
Right: Necklace and earrings set with blocks in gold, 22-inch chain, circa 1980s. $15-25.
Jewelry courtesy of Kenneth L Surratt, Jr.

Metal gold and silver pendant necklace with matching ring, 18-inch chain, circa 1980s. *Courtesy of Mary G. Moon.* $15-25.

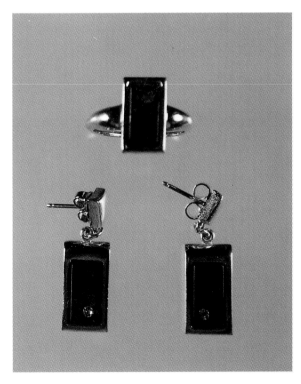

Onyx pierced earrings and ring set with diamonds, circa 1980s. Shown larger than life size. *Courtesy of Kenneth L. Surratt, Jr.* $45-55.

Kenneth Jay Lane Midnight Rose Set, black plastic beads and off-white flower with rhinestones, earrings with rhinestones, 1987. *Courtesy of Miss Mary Jane.* $55-65.

Kenneth Jay Lane Antiques Impressions pin and earrings with rhinestones and pearls, 1990. *Courtesy of Miss Mary Jane.* $65-75.

Rich Christmas Earrings and brooch with enameled leaves, red stone, and pearls, circa 1990s. *Courtesy of Miss Mary Jane.* $20-25.

Pink Tulip Earrings and Pin with pink rhinestones set in silver with floral motif, circa 1990s. *Courtesy of Miss Mary Jane.* $25-35.

Top: Silver hoops with bells pierced earrings, circa 1990s. $10-15.
Bottom: Avon Christmas Tree Pin with red stones, circa 1990s. $15-25.
Jewelry courtesy of Miss Mary Jane.

Gold with pearl pin and pierced earrings set, circa 1990s. *Courtesy of Miss Mary Jane.* $25-35.

Bee pin and earrings set with gold and pearl earrings and pin with gold, pearl, and amethyst stones, circa 1990s. *Courtesy of Miss Mary Jane.* $25-35.

Heart pin and earrings set with pearl, circa 1990s. *Courtesy of Miss Mary Jane.* $25-35.

Ivory and black enameled silver butterfly pin and pierced earrings, circa 1990s. *Courtesy of Miss Mary Jane.* $25-35.

Ivory and black enameled silver fan-shaped pin and pierced earrings, circa 1990s. *Courtesy of Miss Mary Jane.* $25-35.

Pearl necklace with gold butterfly and pearl and gold butterfly earrings, 14-inch necklace, circa 1980s. *Courtesy of Miss Mary Jane.* $35-45.

Necklace and earrings set, necklace with colored beads and purple and gold earrings, circa 1990s. *Courtesy of Miss Mary Jane.* $18-22.

Enameled mauve and gold butterfly pin and pierced earrings with rhinestones, circa 1990s. *Courtesy of Miss Mary Jane.* $30-40.

Fluttering Hummingbird Pin and Just for Spring Hummingbird Earrings, circa 1990s. *Courtesy of Miss Mary Jane.* $25-35.

Plastic red and black 18-inch necklace and earrings set, circa 1980s. *Courtesy of Miss Mary Jane.* $20-25.

Elegant Gift Set with gold bracelet and earrings with amethyst colored stones, circa 1990s. *Courtesy of Miss Mary Jane.* $25-35.

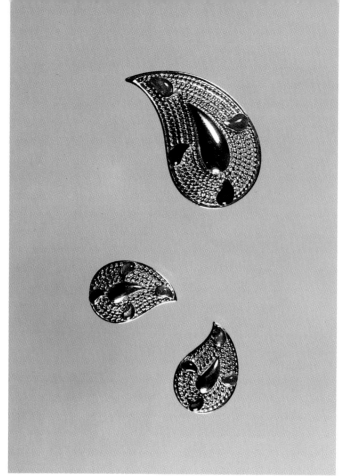

Gold pin and earrings set with colored stones, circa 1990s. *Courtesy of Miss Mary Jane.* $25-35.

Gold Ornate Cross Earrings and Necklace with pearls, circa 1990s. *Courtesy of Miss Mary Jane.* $25-35.

Enameled poinsettia pin/pendant and earrings set, circa 1980s. *Courtesy of Miss Mary Jane.* $20-25.

Gold pierced earrings and pendant necklace with cat and heart motif, 26-inch necklace, circa 1990s. *Courtesy of Sarah Jones.* $25-35.

Classic Baroque Cream pearl bracelet, pierced earrings, and two necklaces, necklaces 21 and 24 inches, 1982. *Courtesy of Sarah Jones.* $55-65.

Lapis Blue Reflections bracelet, necklace, and interchangeable earrings, 1985. *Courtesy of Sarah Jones.* $45-55.

Chapter Four
Earrings

Gold clip-on Casual Hoop Earrings with seed pearl balls, circa 1990s. *Courtesy of Kenneth L. Surratt, Jr.* $15-20.

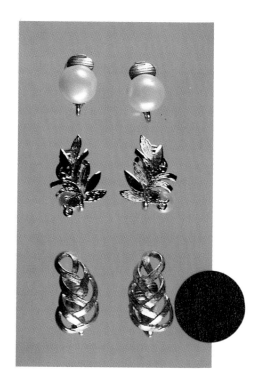

Top: Pearl clip-on earrings, circa 1980s. $10-15.
Center: Silver leaves with pearl clip-on earrings, circa 1980s. $8-12.
Bottom: Gold clip-on earrings, circa 1990s. $10-12.
Earrings courtesy of Kenneth L. Surratt, Jr.

Top: Mistletoe earrings with package, circa 1980s. $10-12.
Center left: Christmas tree earrings, circa 1980s. $10-12.
Center: Christmas package earrings, circa 1980s. $10-12.
Center right: Christmas wreath pin, circa 1980s. $10-15.
Bottom: Enameled poinsettia earrings, circa 1980s. $10-12.
Jewelry courtesy of Virginia Young.

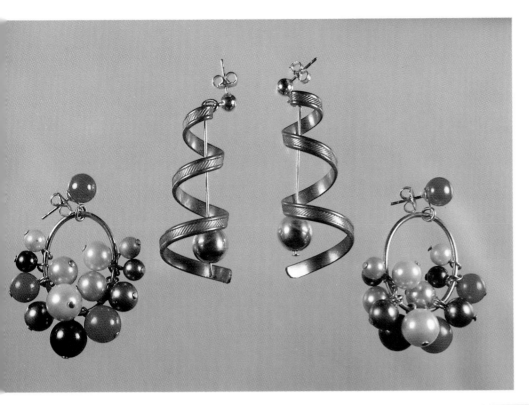

Center: Hot pink Spiral Confetti Earrings with ball, circa 1990s. $12-15.
Left and right: Earrings with multi-colored balls on loop attached to stud, circa 1990s. $10-15.
Earrings courtesy of Kenneth L. Surratt, Jr..

Left: Gold earrings, circa 1980s. $8-12.
Center: Gold earrings with pearls, circa 1990s. $15-20.
Right: Gold earrings, circa 1990s. $15-18.

Silver clip-on earrings with dangling hoops, circa 1980s. $10-15.

Left: Earrings with birds set with turquoise stone, circa 1980s. $15-20.
Top: Pearl clip-on earrings, circa 1980s. $12-18.
Bottom center: Gold clip-on earrings with white stones, circa 1990s. $10-15.
Right: Avon Heart of America Earrings with enameled stars and stripes, circa 1990s. $12-18.

Top: Snowflake gold clip-on earrings, circa 1980s. $10-15.
Bottom: Gold dangle clip-on earrings, circa 1980s. $10-15.

Top: Avon Romantic Splendor Pierced Earrings with amethyst colored stone and rhinestones, circa 1990s. *Courtesy of Kenneth L. Surratt, Jr.* $20-25.
Bottom: Winged Flight Earrings with lapis stone dangles, circa 1990s. *Courtesy of Mary G. Moon.* $20-25.

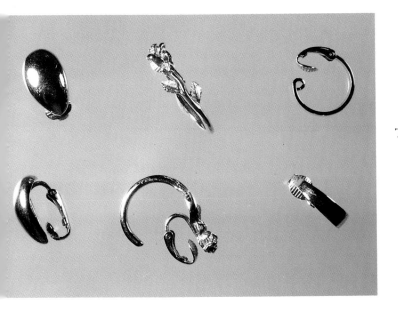

Three pair of gold clip-on earrings, circa 1980s. $15-20.

Top left: Gold Christmas wreath pierced earrings, circa 1990s. $9-12.
Bottom left: Gold pierced earrings with ruby red stones, circa 1980s. $10-15.
Top center: Jade and gold earrings, circa 1980s. $12-15.
Bottom center: Gold pierced earrings, circa 1990s. $10-15.
Right: Gold pierced earrings from the Plaza IV Collection with amethyst stone drops, 1976. $15-20.

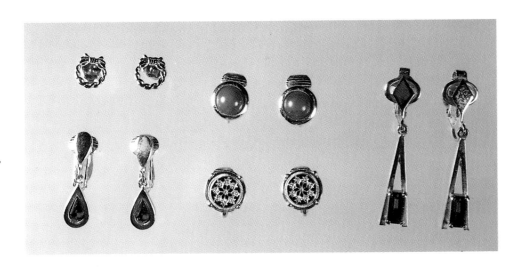

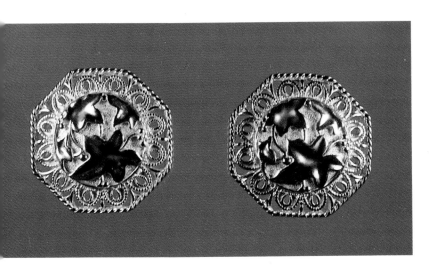

Silver clip-on earrings, circa 1980s. *Courtesy of Kenneth L. Surratt, Jr.* $15-18.

Top left: Drop anchor gold pierced earrings ears, 1992. $10-15.
Bottom left: Spyro Gyro gold tone with pearl pierced earrings, 1993. $10-15.
Right: Color Fun Dangle Pierced Earrings, circa 1990s. $10-15.
Earrings courtesy of Kenneth L. Surratt, Jr.

Top left: Pearly Heart Pierced Earrings with pearls, circa 1990s. $10-15.
Top right: Gold studs with white plastic petals, circa 1980s. $9-12.
Center: Red plastic petals that can replace white ones on stud in *Top right:* $10-15.
Bottom left: Gold basket earrings with pearls, circa 1990s. $12-15.
Bottom right: Sparkle Star Pierced Earrings with rhinestones, circa 1990s. $10-15.
Earrings courtesy of Kenneth L. Surratt, Jr.

Gold heart earrings in red heart box, 1985.
Courtesy of Virginia Young.
$15-18.

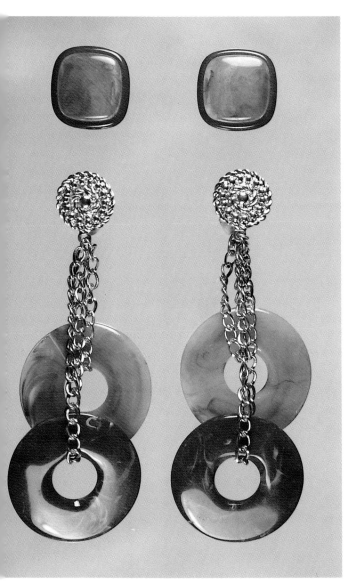

Top: Amber colored plastic earrings, circa 1970s. $9-12.
Bottom: Color Disc Pierced Earrings with large plastic amber colored dangles, circa 1980s. $20-30.
Earrings courtesy of Kenneth L. Surratt, Jr.

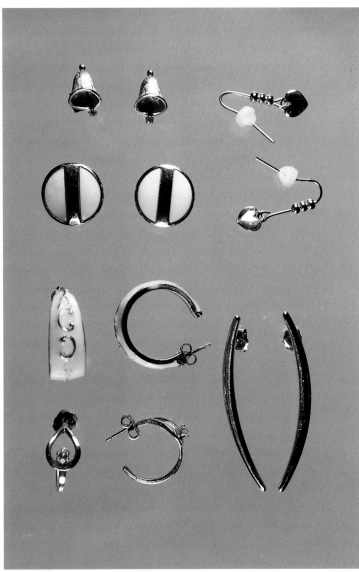

Avon earrings in a variety of styles, circa 1980s-1990s. $15-20 per pair.

Left: Gold pierced earrings with lavender plastic, circa 1980s. $8-12.
Center: Gold earrings with turquoise plastic beads and gold dangles, circa 1990s. $10-15.
Right: Gold disc earrings with geometric design, circa 1990s. $10-15.
Earrings courtesy of Kenneth L. Surratt, Jr.

Top: Sterling pierced earrings with zodiac sign of Cancer with pearl, 1996. $15-20.
Bottom: Gala night's clip-on rhinestone earrings with silver backs, 1991. $15-20.
Earrings courtesy of Kenneth L. Surratt, Jr.

Left: Gold flower pin, 1971. $15-20.
Right: Gold earrings with royal blue enameling for pierced ears, circa 1990s. $15-20.
Jewelry courtesy of Kenneth L. Surratt, Jr.

Top left: Gold promotional incentive earrings, 1986. $15-20.
Top right: Earrings with clock face, 1986. $20-25.
Bottom: Earrings with light sapphire stones, 1975. $20-25.
Earrings courtesy of Kenneth L. Surratt, Jr.

Left: Gold necklace with lapis stone, circa 1990s. $20-15.
Center left: Gold spiral pierced earrings, circa 1980s. $15-20.
Center right: Two gold lapel pins with letter "S", circa 1990s. $20-30.
Right: Gold earrings, 1971. $15-20.
Jewelry courtesy of Kenneth L. Surratt, Jr.

Trinket box that was part of a set with a bud vase and picture frame to match (not pictured) to commemorate Mother's Day 1986. Contains children's jack earrings of yellow and blue plastic, 1986. $10-15 each.
Items courtesy of Virginia Young.

Avon gold sun earrings, circa 1990s. *Courtesy of Kim Lightfoot.* $15-25.

Blue earrings with interchangeable magnetic parts, circa 1980s. *Courtesy of Virginia Young.* $15-25.

Top: Aqua-colored Metallic Spiral Earrings, circa 1990s. $10-12.
Bottom: Clip-on gold turquoise and orange colored Graphic Earrings, 1992. $15-18.
Earrings courtesy of Kenneth L. Surratt, Jr.

Top: True to the Heart earrings with garnet red stones set in gold, 1992. $9-12.
Center: Sweet Mesh Earrings dangles with pearl, 1994. $15-20.
Bottom: Precious Brilliance knot pierced earrings with rhinestones, 1996. $15-20.
Earrings courtesy of Kim Lightfoot.

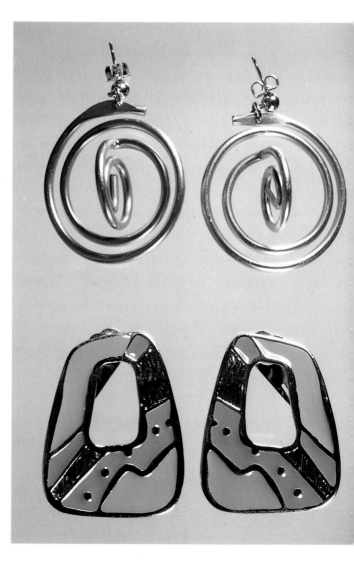

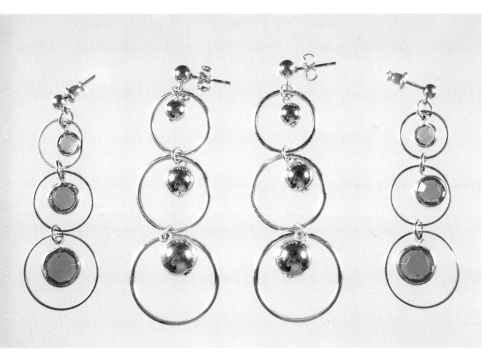

Left and right. Fashion Lever Back Earrings with pink plastic petals and pearls, 1995. $12-16.
Center: Swinging Staple Gold Stone dangle earrings, circa 1990s. $10-15.
Earrings courtesy of Kenneth L. Surratt, Jr.

Top: Tailored initial earrings, 1992. $10-15.
Center: Cheery Cherry Pierced Earrings made of gold and glass, 1992. $12-18.
Bottom: Spring Treasures pierced earrings of gold with flower design, 1992. $15-18.
Earrings courtesy of Kenneth L. Surratt, Jr.

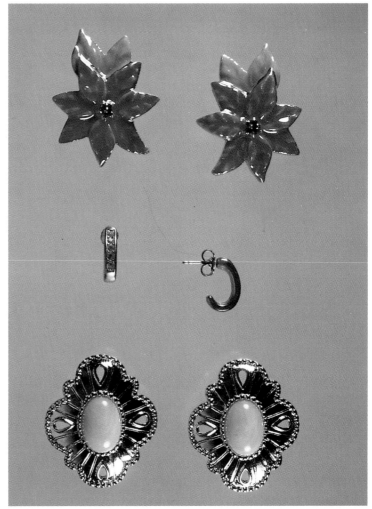

Top: Porcelain Poinsettia Pierced Earrings of red and green glass, 1992. $12-15.
Center: Small gold pierced earrings with rhinestones, circa 1990s. $10-15.
Bottom: Silvery Sands Pierced Earrings silver plated with pink cabochon, 1990. $15-18.
Earrings courtesy of Kenneth L. Surratt, Jr.

Top: Gold Tailored Convertible Earrings with Removable Hoops, circa 1990s. $18-22.
Bottom: Pearlized earrings with gold, circa 1980s. $15-20.
Earrings courtesy of Geraldine Heather.

Gold earrings with enameled red, white, and blue, circa 1990s. *Courtesy of Miss Mary Jane.* $15-20.

Festive Swivel Earrings with steel posts, 1995. *Courtesy of Geraldine Heather.* $15-18.

Left and right. Spring Swing earrings with pearls, plastic leaves, and plastic pink and violet flowers, 1995. $15-18.
Center: Gold dangle earrings with links, 1991. $15-20.
Earrings courtesy of Kenneth L. Surratt, Jr.

Top: Avon Angelic Expressions Clip-On Earrings with pearl rhinestone angel and halo, circa 1990s. $15-20.
Center: Gold clip-on earrings, 1977. $15-18.
Bottom: Gold and pearl clip-on earrings with rhinestones, circa 1980s. $15-20.
Earrings courtesy of Geraldine Heather.

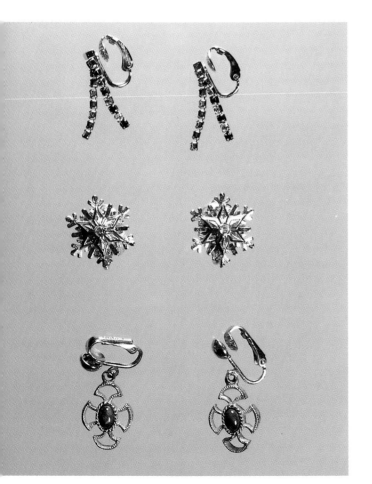

Top: Dangle clip-on earrings with rhinestones and emerald green stones, circa 1990s. $15-18.
Center: Gold with rhinestones snowflake earrings, circa 1980s. $15-20.
Bottom: Clip-on earrings with jade stones, circa 1970s. $15-18.
Earrings courtesy of Geraldine Heather.

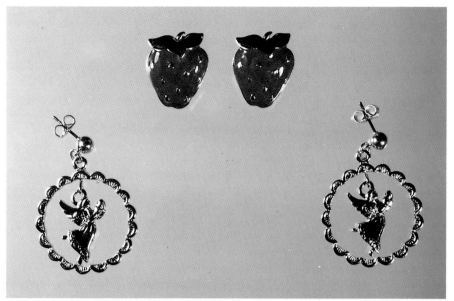

Top: Strawberry pierced earrings, circa 1990s. $12-18.
Bottom: Soaring Angel Pierced Earrings, 1992. $15-18.
Earrings courtesy of Donna Smith.

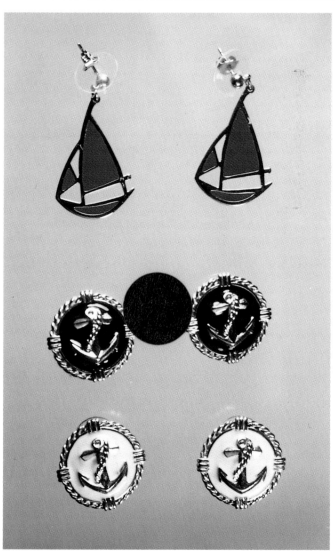

Top: Red enamel sailboat earrings for pierced ears, circa 1990s. $15-20.
Center: Navy enameled nautical earrings with gold anchor, circa 1990s. $15-25.
Bottom: White enameled nautical earrings with gold anchor, circa 1990s. $15-25.
Earrings courtesy of Sarah Jones.

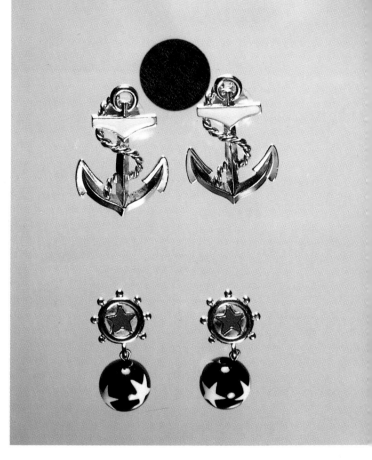

Top: Nautical pierced earrings of gold with white enamel, circa 1990s. $15-25.
Bottom: Star Spangle Dangle Pierced Earrings with red stars and blue balls with white stars, circa 1990s. $15-20.
Earrings courtesy of Sarah Jones.

Top left: Delicate Drop Pierced Earrings with gold and pink glass with dangles, circa 1990s. *Courtesy of Kim Lightfoot.* $15-20.
Top right: Special Heart Pierced Earrings with seed pearls, circa 1990s. *Courtesy of Kenneth L. Surratt, Jr.* $18-22.
Bottom center: Pearlized shell shaped pierced earrings with gold, circa 1990s. *Courtesy of Kenneth L. Surratt, Jr.* $10-15.

Geometric gold dangle pierced earrings, circa 1990s. *Courtesy of Kim Lightfoot.* $15-18.

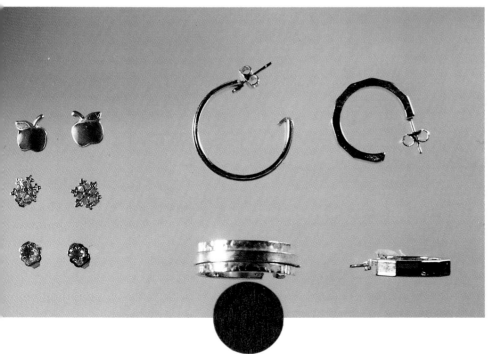

Left top. Red and green enameled apple pierced earrings, circa 1980s. $15-18.
Left center: Gold snowflake pierced earrings with rhinestones, circa 1980s. $15-20.
Left bottom: Pierced earrings with ruby red stone, circa 1980s. $15-20.
Right. Two pair of silver pierced earrings, circa 1980s. $15-25 each.
Earrings courtesy of Mary Jo Michaelis.

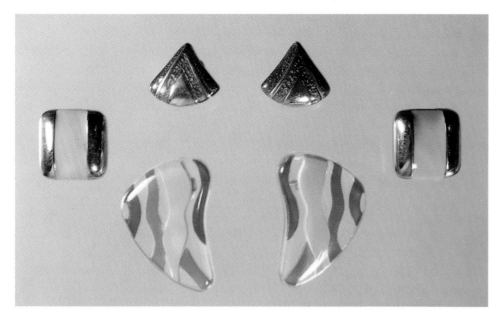

Top: Green and gold enameled earrings, circa 1980s. $9-12.
Left and right. Gold and plastic pierced earrings, circa 1980s. $10-15.
Bottom center: Lucite clear pierced earrings with green and turquoise, circa 1980s. $9-15.
Earrings courtesy of Mary G. Moon.

Top: Gold clip-on earrings, circa 1980s. $12-15.
Center: Plastic orange and blue pierced earrings, circa 1980s. $8-12.
Bottom: White plastic clip-on earrings, circa 1980s. $9-12.
Earrings courtesy of Mary G. Moon.

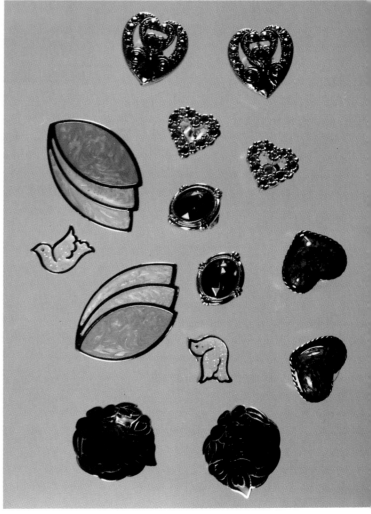

Top: Heart shaped silver pierced earrings, circa 1990s. $20-25.
Top center: Heart shaped pierced earrings with amethysts, circa 1990s. $18-22.
Center left: Enameled pierced earrings, circa 1980s. $15-18.
Center left: Yellow enameled bird earrings, circa 1980s. $12-15.
Center: Amethyst pierced earrings, circa 1990s. $25-35.
Center right: Enameled heart pierced earrings, circa 1980s. $12-15.
Bottom: Black and gold enameled pierced earrings, circa 1980s. $12-15.
Earrings courtesy of Miss Mary Jane.

Chapter Five
Rings

Top left: Ring with pink and clear stones, circa 1970s. $35-45.
Top right: Ring with clear stone, circa 1970s. $35-45.
Center left: Ring with diamond, circa 1970s. $45-55.
Center: Pearl ring glacé, circa 1970s. $25-35.
Center right: Ring with agate stone, circa 1970s. $45-55.
Bottom left: Ring with clear stone, circa 1970s. $25-35.
Bottom center: Domed gold ring, circa 1970s. $30-40.
Bottom right: Jade and gold ring, circa 1970s. $25-35.

Avon ring with amethyst stones, circa 1980s. Photographed larger than life size. $44-55.

Top: Silver metal with garnet ring, circa 1970s. $35-45.
Center left: Silver metal ring with clear stone, circa 1970s. $30-40.
Center: Silver metal turquoise ring, circa 1970s. $35-45.
Center right: Silver ring, circa 1970s. $25-35.
Bottom: Sterling silver key in shape of ring, circa 1970s. $25-35.

Top left: Gold buckle ring with rhinestones, circa 1980s. $35-45.
Top right: Gold ring with garnet stones, circa 1980s. $45-55.
Center left: Ring with onyx stone, circa 1980s. $55-65.
Center right: Gold ring with rhinestones and emerald green stones, circa 1970s. $35-45.
Bottom left: Gold ring, circa 1980s. $35-45.
Bottom right: Gold ring, 1973. $35-45.
Rings courtesy of Kenneth L. Surratt, Jr.

Magnified view of Avon gold-plated ring with onyx surrounded by seed pearls, circa 1970s. $45-55.

Top left: Ring with rhinestones and pearl, circa 1980s. $35-45.
Top right: Silver ring, circa 1980s. $25-35.
Center left: Mother-of-pearl ring, circa 1970s. $35-45.
Center right: Gold ring, circa 1990s. $25-35.
Bottom left: Midnight Splendor Ring with onyx stones and rhinestones, 1973. $45-55.
Bottom right: Sterling silver ring, circa 1990s. $35-45.
Rings courtesy of Kenneth L. Surratt, Jr.

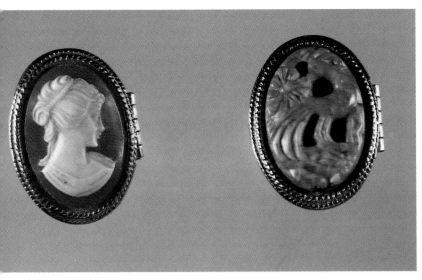

Left: Plastic cameo ring with glacé, circa 1980s. $20-25.
Right: Jade-colored plastic ring with glacé, circa 1980s. $18-22.
Both items photographed larger than life size.
Jewelry courtesy of Kenneth L. Surratt, Jr.

Top left: Gold ring with pearls and ruby red stone, circa 1980s. $20-25.
Center: Evening Creation Cluster Ring, one pearl missing, 1972. $25-30 as is.
Top right: Gold ring with pearl, circa 1970s. $20-25 as is.

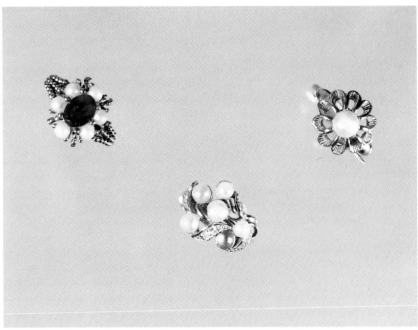

Left: Gold and silver metal ring, circa 1980s. $25-35.
Bottom center: Silver ring with clear stones, circa 1980s. $40-50.
Right: Ring with onyx and small diamond, circa 1970s. $45-55.
Rings courtesy of Mary G. Moon.

Chapter Six
Pins

Pin with floral design, circa 1980s. $25-35.

Left: Gold brooch/pendant with pale blue pearl and eight aquamarine blue stones, circa 1980s. $35-45.
Right: Star-shaped rhinestone pin, circa 1980s. $25-35.
Jewelry courtesy of Kenneth L. Surratt, Jr.

Top: Avon heart brooch with ruby red stone on bow, circa 1980s. $25-35.
Bottom: Bear pin with rhinestones, circa 1980s. $30-40.
Jewelry courtesy of Mary G. Moon.

Brooch depicting mask of gold with red, purple, and black stones, circa 1980s. $45-55.

Left: Starflower Pin, 1972. $35-45.
Right: Kenneth Jay Lane gold brooch with pearls and clear stones, signed K.J.L. for Avon, circa 1980s. *Courtesy of Kenneth L. Surratt, Jr.* $45-55.

Top left: Basket lapel pin, circa 1980s. $15-20.
Top right: Gold lapel pin in shape of Texas and inscribed "TEXAS", circa 1990s. $15-20.
Bottom left: Gold scissors lapel pin inscribed "Hairdresser", circa 1990s. $15-25.
Bottom center: Frog lapel pin with green enameling, circa 1980s. $15-20.
Bottom right: Holy Family lapel pin with navy blue enamel, circa 1990s. $15-25.

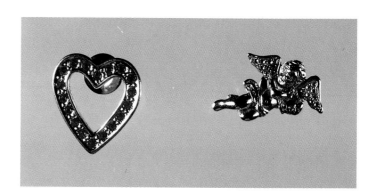

Left: Faux Rubies Sparkling Heart Pin with ruby red stones, circa 1990s. $25-35.
Right: Silver cherub lapel pin, circa 1990s. $15-25.

Top: Crown pin with rhinestones, circa 1980s. $25-35.
Bottom: Avon Beautiful Butterfly Pin with pearls, circa 1990s. $20-30.
Pins courtesy of Kenneth L. Surratt, Jr.

Left: Silver Avon award key stickpin, circa 1990s. $10-15.
Right: Gold stickpin with koala bears, circa 1990s. $20-30.

Gold brooch with plastic grape clusters, circa 1970s. $12-18.

Left: Gold turtle pin with multi-colored stones, circa 1980s. $20-25.
Center: Copper flower pot pin with silk flowers, circa 1990s. *Courtesy of Kenneth L. Surratt, Jr.* $15-20.
Right: Gold and coral plastic flower pin, circa 1990s. *Courtesy of Kenneth L. Surratt, Jr.* $15-25.

Two plastic flower pins with seed pearl center, green leaves, and stem, circa 1980s. $15-25.

Pin with silk rose, 4 inches, circa 1980s. $15-18.

Flower brooch with gold and white enamel, circa 1980s. $25-35.

Bow and heart pin inscribed, "I Love You Mother", circa 1980s. *Courtesy of Lila Heather.* $25-35.

Left: Avon copper metal bow pin with silk flowers, 1980. $20-25.
Right: Silver metal branch shaped pin, circa 1980s. *Courtesy of Mary G. Moon.* $18-22.

Left: Gilded bird lapel pin, 1976. $25-35.
Center left: Unicorn stickpin, circa 1990s. $18-22.
Center right: Bar pin with rhinestone, circa 1980s. $15-20.
Right: Heart-shaped lapel pin, circa 1980s. $15-18.
Jewelry courtesy of Donna Smith.

Gold nautical pin with red, white, and blue enamel, circa 1990s. *Courtesy of Sarah Jones.* $25-35.

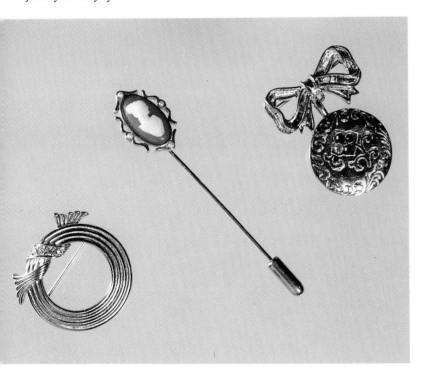

Sparkling Heart Pin with Rhinestones, circa 1990s. *Courtesy of Geraldine Heather.* $15-20.

Left: Gold circular pin with rhinestones, circa 1980s. $18-22.
Center: Cameo stickpin, circa 1990s. $15-20.
Right: Gold pin with bow and four multicolored stones, "DEAR" inscribed on back of round piece, circa 1990s. $25-35.
Pins courtesy of Mary G. Moon.

Left: Stickpin with stars, circa 1980s. $15-20.
Right: Gold Christmas Tree Pin with rhinestone, 1985. $22-25.
Pins courtesy of Mary Jo Michealis.

Left: Ceramic pin with floral motif, circa 1980s. $15-20.
Top center: Siamese cat ceramic pin, circa 1980s. $15-20.
Right: Pansy ceramic pin, circa 1980s. $15-20.
Pins courtesy of Miss Mary Jane.

Top: Gold bow pin with rhinestones, circa 1990s. $25-35.
Bottom: Kenneth Jay Lane black enameled and gold pin, signed K.J.L. for Avon, circa 1990s. $45-55.
Pins courtesy of Miss Mary Jane

Kenneth Jay Lane Nature's Treasures pin with enameled wings and pink stone and rhinestones, 1990. *Courtesy of Miss Mary Jane.* $35-45.

Top left: Gold initial "a" pin, circa 1990s. $15-20.
Top right. Pin inscribed with initial "M", circa 1980s. $18-22.
Center: "M" pin, circa 1990s. $15-20.
Bottom left: Gold initial "M" pin, circa 1990s. $15-20.
Bottom right: Pendant with initial "M" and charm, circa 1980s. $20-25.
Pins and pendant courtesy of Miss Mary Jane.

Kenneth Jay Lane Antiques Impressions Pin with rhinestone and pearls, 1990. *Courtesy of Miss Mary Jane.* $35-45.

Gold heart with raised silver, circa 1990s. *Courtesy of Miss Mary Jane.* $25-30.

Top: Gold star pin with pearls and blue stones, 1995. $20-30.
Bottom: Gold pin with pearl, circa 1990s. $25-35.
Pins courtesy of Miss Mary Jane.

Silent Night Colors of Christmas Ceramic Pin, circa 1980s. *Courtesy of Miss Mary Jane.* $15-20.

Pin with pearls, circa 1970s. *Courtesy of Miss Mary Jane.* $25-35.

Top: Small Juliette Gordon Low swallow from Smithsonian Institution series, circa 1990s. $25-35.
Bottom left: Pearlized pin with silver, circa 1980s. $20-30.
Bottom right: Pin with colored plastic stones, circa 1980s. $20-25.
Pins courtesy of Miss Mary Jane.

Top left: Gold arrow stickpin, circa 1980s. $15-20.
Top center: Gold tie clip, circa 1980s. $20-25.
Top right: Tulip stickpin, circa 1980s. $20-30.
Bottom left: Telephone stickpin, circa 1980s. $25-35.
Bottom center: Stickpin from the Pavé Collection with rhinestones, 1984. $25-35.
Bottom right: Tennis racket pin with pearl, circa 1980s. $20-30.
Pins courtesy of Miss Mary Jane.

Gold apple pin, circa 1980s. $15-20.

Left: Cross pin/pendant with topaz stone, circa 1990s. $35-45.
Right: Beautiful Butterfly Pin with pink stones, blue stones, and pearls, circa 1990s. $35-45.
Pins courtesy of Miss Mary Jane.

Chapter Seven
Bracelets

José Mariá Barrera hammered metal Corinthian bracelet with box, 1989. *Courtesy of Sarah Jones.* $35-45.

Men's identification bracelet, circa 1980s. $35-45.
Two men's tie tacks, 1985. $15-20 each.
Items courtesy of Virginia Young.

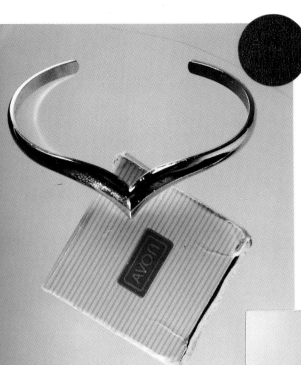

Avon silver cuff bracelet, circa 1990s. $20-25.

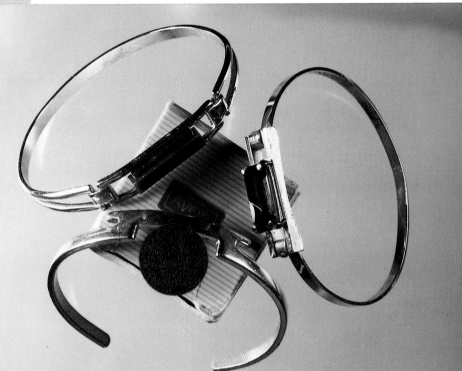

Top left: Gold cuff bracelet, circa 1990s. *Courtesy of Mary G. Moon.* $25-35.
Top right: Gold bracelet with jade stone, circa 1990s. $30-35.
Bottom: Gold bracelet with amethyst stone, circa 1990s. $30-35.

Left: Avon cuff bracelet inscribed, "Gemini II—imaginative alert witty versatile", circa 1990s. $25-35.
Right: Avon cuff bracelet inscribed, "Aquarius—progressive original intelligent sincere", circa 1990s. $25-35.

Avon twisted gold bracelet, circa 1990s. $35-45.

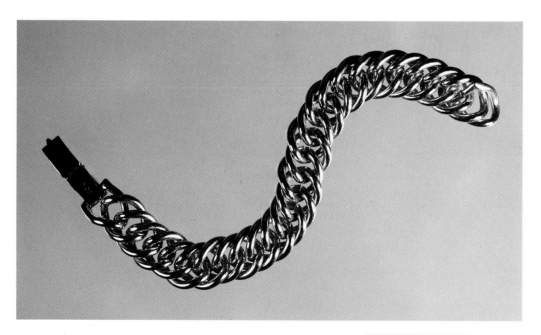

Avon gold bracelet, circa 1990s. *Courtesy of Kenneth L. Surratt, Jr.* $25-35

Avon gold bracelet, circa 1990s. $35-45.

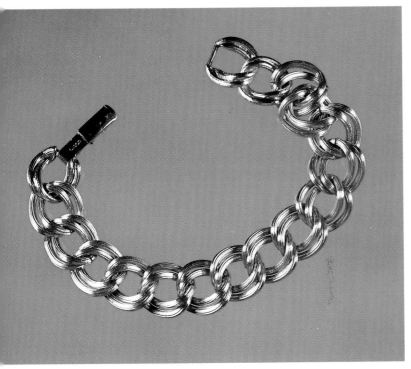

Avon gold chain bracelet, circa 1990s. *Courtesy of Kenneth L. Surratt, Jr.* $25-35.

Avon bracelet with colored plastic, circa 1990s. *Courtesy of Mary G. Moon.* $25-35.

Avon gold cuff bracelet with heart locket, 1993. *Courtesy of Kenneth L. Surratt, Jr.* $35-45.

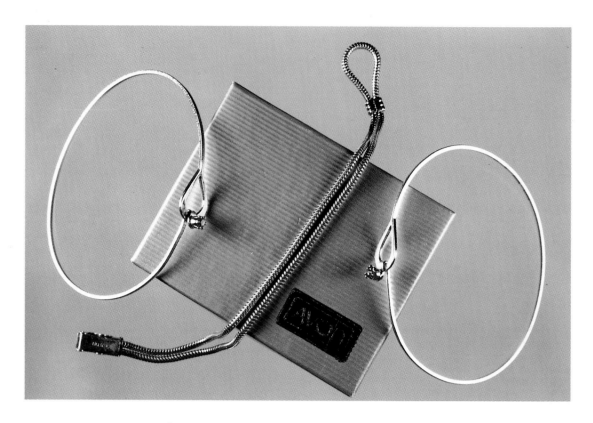

Left: Avon gold bracelet with topaz stone, circa 1980s. $35-45.
Center: Gold bracelet with two heavy chains, circa 1980s. $25-35.
Right: Gold bracelet with ruby red stone, circa 1980s. $35-45.
Bracelets courtesy of Kenneth L. Surratt, Jr.

Gold cuff Leo zodiac bracelet, circa 1990s.
Courtesy of Kenneth L. Surratt, Jr. $25-35.

Avon gold bangle bracelet with violet exterior, circa 1990s. *Courtesy of Kenneth L. Surratt, Jr.* $25-35.

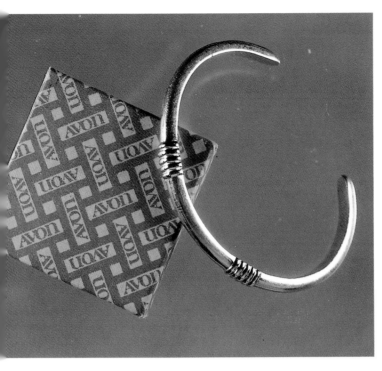

Avon gold cuff bracelet, circa 1990s. *Courtesy of Kenneth L. Surratt, Jr.* $35-45.

Avon gold ankle bracelet, circa 1980s. $20-30.

Gold bracelet with lapis cabochon, circa 1990s. *Courtesy of Kenneth L. Surratt, Jr.* $35-45.

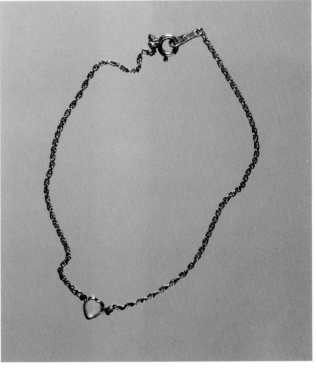

Avon silver ankle bracelet, circa 1980s. $20-25.

Avon gold bracelet with pearl, circa 1990s. *Courtesy of Kim Gibson.* $45-55.

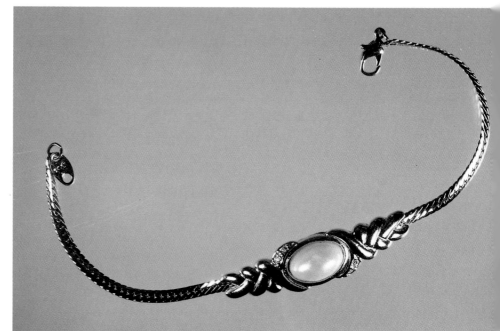

Avon tennis bracelet with rhinestones, 7 inches, circa 1990s. $35-45.

Avon gold bangle, 3 inches in diameter, circa 1980s. $35-45.

Gold bracelet with porcelain butterfly, circa 1990s. *Courtesy of Libby Warren.* $25-35.

Gold bracelet with mesh, circa 1990s. *Courtesy of Donna Smith.* $35-45.

Left: Watch face with cat's portrait and dark leather watch band, circa 1990s. $35-45.
Right: Watch with multi-colored face and white leather band, circa 1990s. $45-55.
Watches *courtesy of Sarah Jones.*

Left: Watch with gold link band, circa 1980s. $35-45.
Right: Heart shaped watch with rhinestones, circa 1990s. $40-50.
Watches courtesy of Sarah Jones.

Avon Award watch with assorted watch band wardrobe, circa 1990s. *Courtesy of Sarah Jones.* $65-85.

Left: Silver bracelet with locket and initials, circa 1980s. $35-45.
Right: Bracelet with clear stone, circa 1980s. $25-35.
Bracelets courtesy of Mary Jo Michaelis.

Left: Tennis bracelet with rhinestones and pearls, circa 1990s. $30-40.
Center: Tennis bracelet with ruby red stones, circa 1990s. $35-45.
Right: Tennis bracelet with rhinestones, circa 1990s. $35-45.
Tennis bracelets courtesy of Miss Mary Jane.

Chapter Eight
Necklaces

Avon owl pendant, 2.75 x 1 inch, circa 1980s. *Courtesy of Martha A. McDougal.* $30-40.

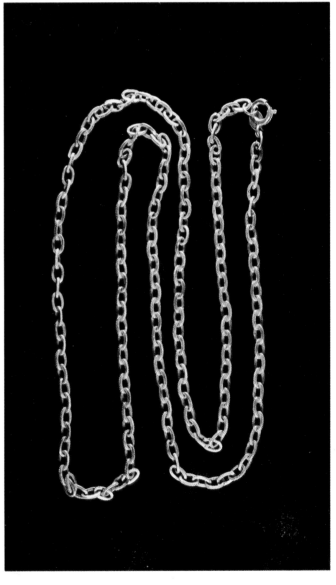

Avon chain, 30 inches, circa 1980s. *Courtesy of Kenneth L. Surratt, Jr.* $18-22.

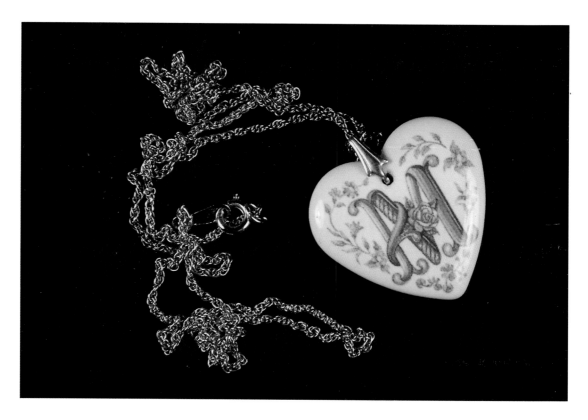

Porcelain heart Mother's Day pendant with "M", 36-inch chain, circa 1980s. *Courtesy of Martha A. McDougal.* $25-35.

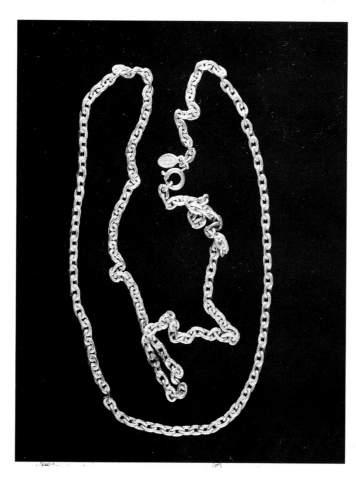

Avon gold-tone chain, 28 inches, circa 1980s. *Courtesy of Kenneth L. Surratt, Jr.* $18-22.

Town and Country pendant necklace, 3-inch pendant, 28-inch chain, circa 1980s. *Courtesy of Kenneth L. Surratt, Jr.* $35-45.

Avon gold chain with six ruby red stones, 44-inch chain, circa 1990s. *Courtesy of Kenneth L. Surratt, Jr.* $20-25.

Gold necklace with three gilded strands and two mother of pearl stones, chains measure 22, 19, and 17 inches, circa 1980s. *Courtesy of Kenneth L. Surratt, Jr.* $25-30.

New Purple Pendant with pearls and tassels, 1972. *Courtesy of Jean Flippo.* $25-35.

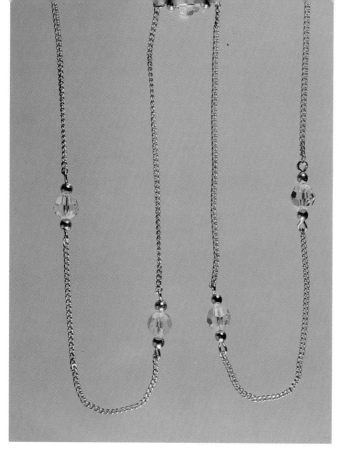

Avon gold necklace with crystals, circa 1980s. *Courtesy of Jean Flippo.* $20-25.

Ebony Teardrop pendant with chain, circa 1980s. $35-45.

Plastic and gold necklace, 24-inch chain, 1977. $35-45.

Gold necklace with pearls, 1971. *Courtesy of Virginia Young.* $35-45.

Elusive Perfume reversible pendant necklace with faux clock face, 31-inch chain, circa 1980s. $35-45.

Reverse side of Elusive Perfume pendant necklace.

Elegant Apple Pendant with rhinestones on gold leaves, 30-inch chain, circa 1990s. *Courtesy of Sarah Jones.* $20-30.

Jade-colored plastic beaded necklace, circa 1990s. $20-30.

Pendant with initial "S" and charms, 24-inch chain, circa 1980s. *Courtesy of Sarah Jones.* $25-35.

Kenneth Jay Lane elephant pendant signed "K.J.L. for Avon", 30-inch chain, circa 1980s. *Courtesy of Sarah Jones.* $55-65.

Left: Gold 36-inch chain, circa 1990s. $20-25.
Middle. Pendant with glacé, 19-inch chain, circa 1970s. $25-35.
Right: Gold 16-inch chain necklace, circa 1990s. $20-25.

Gold pendant with signs of zodiac, 30-inch chain, circa 1980s. $35-45.

Left: Initial "F" pendant lariat necklace with pearl on 24-inch chain, circa 1980s. $20-30.
Center left: Opal pendant necklace, 18-inch chain, circa 1980s. $35-45.
Center: Onyx pendant with small diamond, 18-inch chain, circa 1980s. $45-55.
Center right: Pendant with amethyst stone, 13-inch chain, circa 1990s. $18-22.
Right: Gold block "F" pendant, 18-inch chain, circa 1980s. $15-20.

Left: Gold and silver pendant, 19-inch chain, circa 1980s. $15-20.
Center: Scorpio zodiac pendant necklace, 19-inch chain, circa 198[0s]. $20-30.
Right: Plastic initial "M" pendant necklace, 18-inch chain, circa 1980s. $15-20.

Top left: Silver whistle pendant, 13-inch chain, circa 1980s. $25-35.
Bottom left: Gold and silver butterfly pendant necklace, 14.5-inch chain, circa 1980s. $35-45.
Right: Hinged silver metal pendant, 24-inch chain, circa 1980s. $35-4[5].

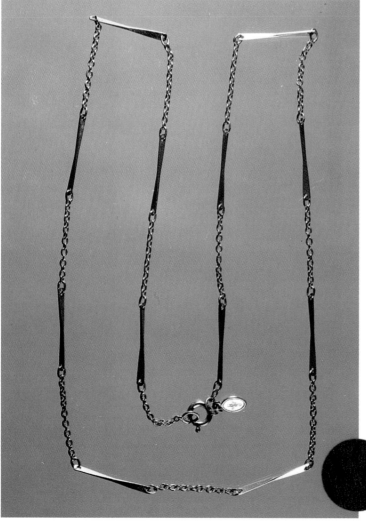

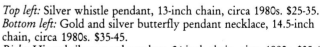

Avon silver metal chain, circa 1980s. $20-25.

Pendant with onyx and pearls on 20-inch chain, circa 1980s. *Courtesy of Doris Vaughan.* $35-45.

Avon necklace with seven amethyst stones with pearls on either side and gold chain between, 36-inch chain, circa 1990s. *Courtesy of Kenneth L. Surratt, Jr.* $35-45.

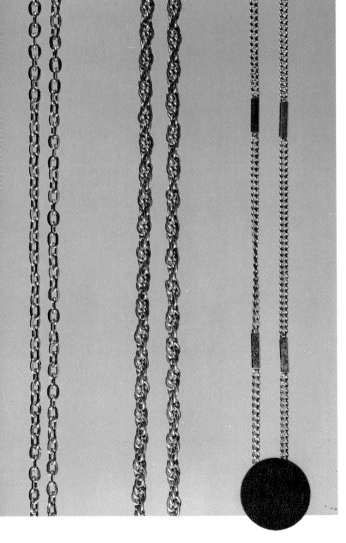

Left: Gold 22-inch chain, circa 1990s. $20-25.
Center: Gold 28-inch chain, circa 1980s. $25-30.
Right: Gold 30-inch chain with gold bars, circa 1980s. $25-30.

Silver pendant with pearl drop, 31-inch chain, circa 1980s. $25-35.

Romantic Rhinestones Velvet Necklace, 13 inches, circa 1980s. *Courtesy of Mary G. Moon.* $15-25.

Heart pendant with clear stones and gold chain, circa 1980s. *Courtesy of Kenneth L. Surratt, Jr.* $25-35.

Gold metal pendant with pink plastic cabochon and seed pearls, gold metal, circa 1980s. *Courtesy of Kenneth L. Surratt, Jr.* $30-35.

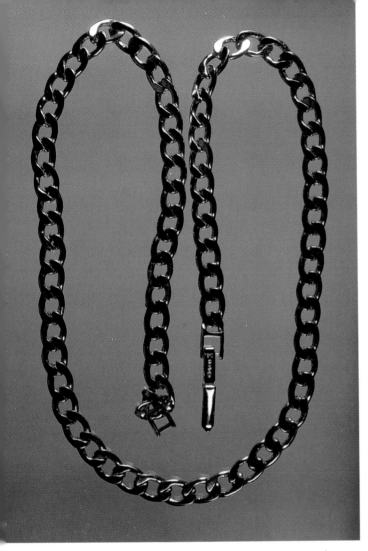

Gold necklace, 19 inches, circa 1980s. *Courtesy of Mary G. Moon.* $15-25.

Topaz stone pendant with glacé, signed "AVON" inside, 37-inch chain, circa 1970s. *Courtesy of Kenneth L. Surratt, Jr.* $35-45.

Left: Green plastic cross with gold trim and gold chain, circa 1980s. $25-35.
Center: Gold cross with ruby red stone and gold chain, circa 1980s. $35-45.
Right: Ivory plastic cross with gold trim and gold chain, circa 1980s. $25-35.
Crosses courtesy of Martha A. McDougal.

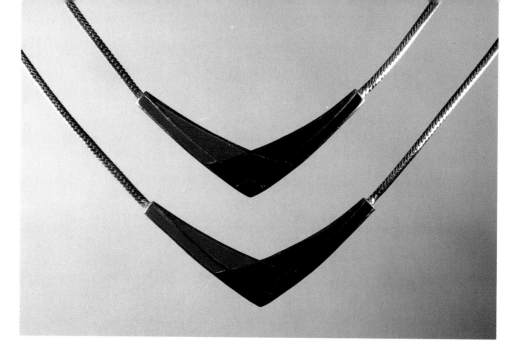

Top: Enameled 18-inch necklace with red and black on gold, circa 1980s. $25-30.
Bottom: Enameled 18-inch necklace with blue and black on gold, circa 1980s. $25-30.
Necklaces courtesy of Kenneth L. Surratt, Jr.

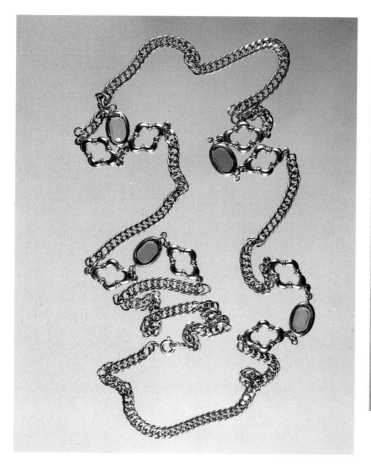

Gold necklace with faceted amethyst stones, 34-inch chain, circa 1980s. *Courtesy of Kenneth L. Surratt, Jr.* $20-30.

Silver 20-inch chain, circa 1990s. *Courtesy of Kenneth L. Surratt, Jr.* $15-25.

Left: Gold pendant from the Plaza IV Collection with amethyst stone and gold chain, 1976. $25-35.
Right: Gold pendant from the Plaza IV Collection with emerald green stone, 1976. $25-35.
Jewelry courtesy of Kenneth L. Surratt, Jr.

Porcelain pendant with hand painted butterfly in blue and yellow, cord for necklace, circa 1980s. *Courtesy of Kenneth L. Surratt, Jr.* $25-35.

Gold fashion bead necklace, 1986. $25-35.

Left: Gold necklace with block letter "M", 18-inch chain, circa 1980s. $15-20.
Top center: Gold necklace with block letter "A", 18-inch chain, circa 1980s. $15-20.
Bottom center: Gold "Mom" necklace with diamond in center of heart, 18-inch chain, circa 1980s. $25-35.
Right: Gold necklace with block letter "K", 18-inch chain, circa 1980s. $15-20.
Necklaces courtesy of Martha A. McDougal.

Enameled gold pendant with floral motif, circa 1980s. $25-35.

Left: Gold 12-inch necklace with five stars, circa 1990s. $15-25.
Center: Gold 16-inch necklace with tassel, circa 1980s. $18-22.
Right: Gold 14-inch necklace with star, circa 1980s. $18-22.
Necklaces courtesy of Martha A. McDougal.

Left: Gold pendant with faceted amethyst stone, gold 16-inch chain, circa 1980s. $35-45.
Center: Gold pendant with onyx stone and groups of seed pearls, gold 20-inch chain, circa 1980s. $40-50.
Right: Pendant with aquamarine stone surrounded by clear stones, gold 12-inch chain, circa 1980s. $25-35.
Necklaces courtesy of Martha A. McDougal.

Left: Gold 18-inch necklace, circa 1980s. $18-22.
Left center: Teardrop necklace with amber stone, 16-inch chain, circa 1980s. $20-30.
Center: Teardrop necklace with clear stone and gold chain, 16-inch chain, circa 1980s. $20-25.
Right center: Opal necklace with gold and gold chain, 16-inch chain, circa 1980s. $25-35.
Right: Necklace with knots, 12 inches, circa 1980s. $20-25.
Necklaces courtesy of Martha A. McDougal.

Left: Gold pendant with clear stone set in incised design, 16-inch chain, circa 1980s. $20-30.
Center: Gold heart pendant with "Mother" engraved, 16-inch chain, circa 1980s. $20-25.
Right: Porcelain pendant with letter "M", gold 18-inch chain, circa 1980s. $20-25.
Necklaces courtesy of Martha A. McDougal.

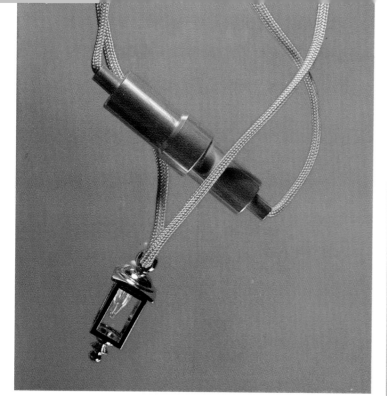

Battery operated light pendant, 10.5-inch rope chain, circa 1990s. $20-25.

Necklace with green and white beads, circa 1980s. *Courtesy of Martha A. McDougal.* $20-30.

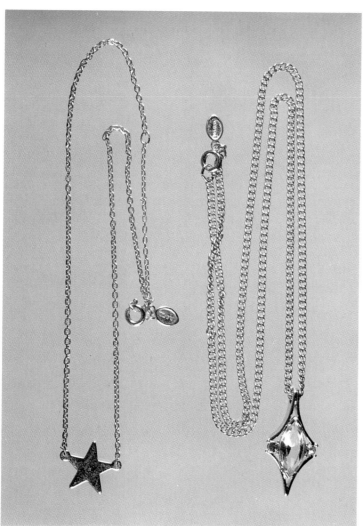

Left: Gold star necklace, 13-inch chain, circa 1980s. $18-22.
Right: Silver pendant with clear stone, silver 20-inch chain, circa 1980s. $20-25.
Necklaces courtesy of Kenneth L. Surratt, Jr.

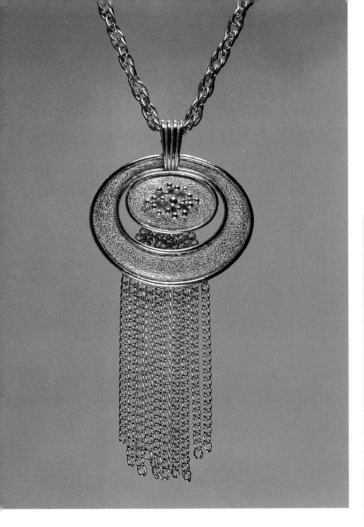

Gold pendant with gold beads and tassel, 33-inch chain, circa 1990s. *Courtesy of Kenneth L. Surratt, Jr.* $25-35.

Gold bead necklace, circa 1980s. $20-25.

Silver with pink cabochon pendant surrounded by rhinestones, with silver chain, 1975. $25-35.

Left: Gold necklace with pink glass flower, 16-inch chain, circa 1980s. $18-22.
Center: Plastic rabbit brooch/pendant with movable legs, 1974. $10-15.
Right: Gold necklace with opaque glass apple and gold leaves, 20-inch chain, circa 1980s. $25-35.
Items courtesy of Martha A. McDougal.

Kenneth Jay Lane butterfly and pearl necklace, signed "K.J.L. for Avon", 17 inches, circa 1980s. $55-65.

White porcelain pendant with hand-painted flowers on 28-inch cord with tassel, circa 1980s. $20-30.

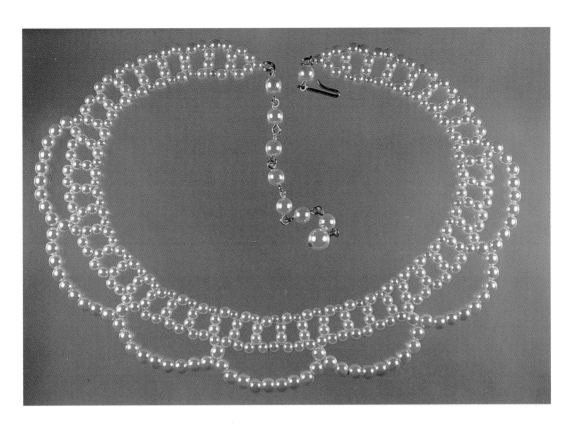

Avon pearlized lace necklace, circa 1970s. *Courtesy of Kenneth L Surratt, Jr.* $45-55.

Necklace with three plastic valentine hearts, circa 1980s. $15-20.

Left: Gold heart necklace, 14-inch chain, circa 1980s. $20-25.
Center: 14K gold Managers Gift Diamond Loop necklace, 1977. $45-55.
Right: Gold necklace with emerald-colored stone surrounded by clear stones, 18-inch chain, circa 1980s. $25-35.
Necklaces courtesy of Kenneth L Surratt, Jr.

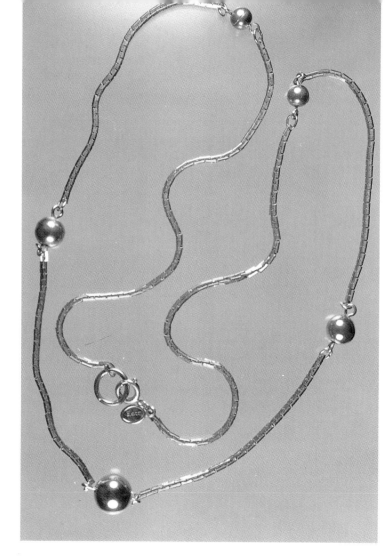

Avon gold 22-inch necklace with five gold balls, circa 1980s. *Courtesy of Kenneth L Surratt, Jr.* $20-25.

Two gold necklaces, each 14 inches long, circa 1990s. *Courtesy of Kenneth L Surratt, Jr.* $15-25 each.

Avon 31-inch chain necklace, circa 1990s. *Courtesy of Kenneth L Surratt, Jr.* $15-25.

Hematite necklace with presentation box and bag, 36 inches, 1987. *Courtesy of Virginia Young.* $45-55.

Necklace with plastic crystals, 54 inches long, circa 1980s. *Courtesy of Virginia Young.* $25-35.

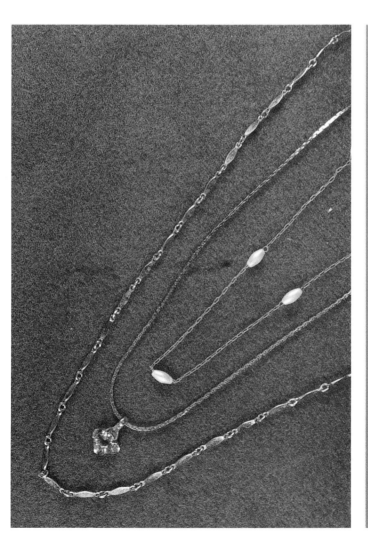

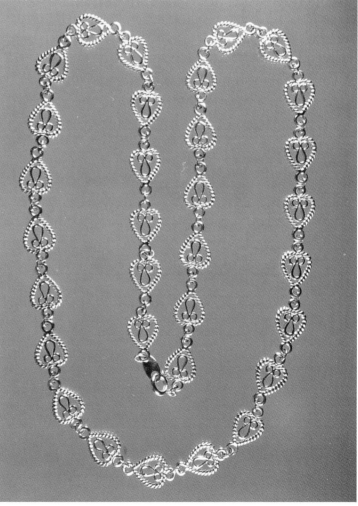

Top: Sterling silver and pearl chain, circa 1980s. $20-25.
Center: Gold chain with amethyst pendant, circa 1980s. $25-35.
Bottom: Gold chain, circa 1980s. $15-20.
Necklaces courtesy of Virginia Young.

Gold Simply Hearts Necklace, 24 iches, 1994. *Courtesy of Kim Lightfoot.* $20-25.

Letter Perfect Necklace with initial "K", 24 inches, circa 1990s. *Courtesy of Kim Lightfoot.* $20-25.

Gold necklace with coral-colored plastic butterfly, 16.5-inch chain, circa 1990s. *Courtesy of Mary G. Moon.* $20-25.

Allure of Silver Sterling silver heart necklace, 32-inch cord, 1992. *Courtesy of Kenneth L. Surratt, Jr.* $20-25.

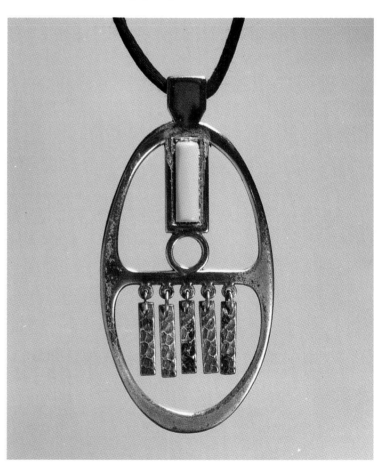

Silver pendant with turquoise plastic, 35-inch black cord, circa 1990s. *Courtesy of Mary G. Moon.* $35-45.

Summer Song The Hummingbird gold and porcelain pendant/brooch, circa 1980s. $45-55. Chain necklace with clusters of seed pearls, 24 inches, circa 1980s. $25-35.

Porcelain pendant depicting blue flowers, 19-inch chain, circa 1980s. *Courtesy of Lila Heather*. $25-35.

Pearl pendant necklace, 18-inch chain, circa 1980s. *Courtesy of Lila Heather*. $25-35.

Plastic cameo with pearl, 25-inch chain with pearls, circa 1970s. *Courtesy of Lila Heather.* $55-65.

Personalized Initial G Locket pendant necklace with initial "G" on enameled white inset, 24-inch chain, circa 1990s. *Courtesy of Geraldine Heather.* $20-25.

Gold and enamel pink flower pendant, 20-inch chain, circa 1980s. *Courtesy of Geraldine Heather.* $25-35.

Avon Romantic Birthstone Pendant with sliding rhinestone, 27-inch chain, circa 1990s. *Courtesy of Lila Heather.* $20-25.

Pink Lace Necklace with porcelain roses and amethyst stone, 18-inch chain, circa 1980s. *Courtesy of Lila Heather.* $20-30.

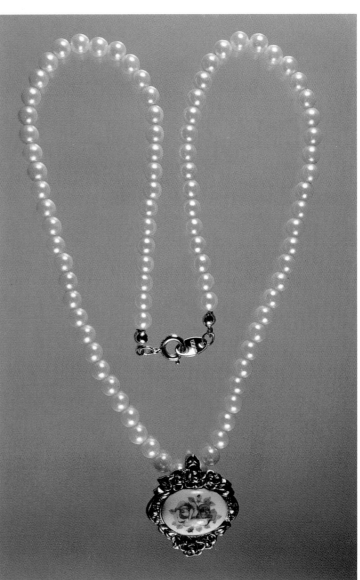

Victorian Romance Porcelain Necklace with 16.5-inch pearl necklace and enameled floral pendant, circa 1980s. *Courtesy of Geraldine Heather.* $25-35.

Heart pendant with enameled flowers, 17-inch chain, circa 1990s. *Courtesy of Lila Heather.* $25-35.

Gold "Mother" pendant with ruby stone, 18-inch chain, circa 1980s. *Courtesy of Geraldine Heather.* $25-35.

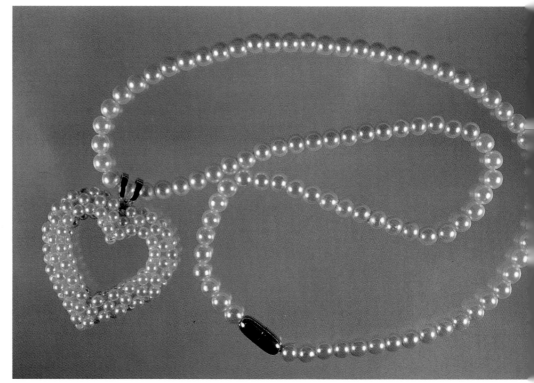

Heart-shaped pearl pendant and 18-inch pearl necklace, circa 1980s. *Courtesy of Geraldine Heather.* $40-50.

Left: Pendant inscribed "Thursday's Child Has Far To Go", 17-inch chain, circa 1980s. $20-25.
Right: "MOM" necklace, 17-inch chain, circa 1980s. $20-25.
Jewelry courtesy of Lila Heather.

Left: Heart pendant with rhinestones, 18-inch chain, circa 1980s. $30-40.
Right: Sparkling Kitty Pendant with rhinestones, 18-inch chain, circa 1990s. $20-25.
Jewelry courtesy of Geraldine Heather.

Pearl and gold heart, 29-inch chain, circa 1980s. *Courtesy of Geraldine Heather.* $25-35.

Romanesque Necklace with colored plastic stones, 12-inch chain, 1976. *Courtesy of Geraldine Heather.* $35-45.

Pendant with turquoise stones and onyx stone in center, circa 1970s. *Courtesy of Geraldine Heather.* $35-45.

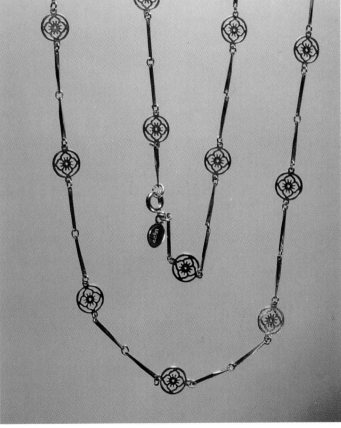

Gold 25-inch necklace with medallions, circa 1990s. *Courtesy of Donna Smith.* $20-30.

Multi-colored 19-inch necklace with plastic and pearl beads, circa 1990s. *Courtesy of Sarah Jones.* $25-35.

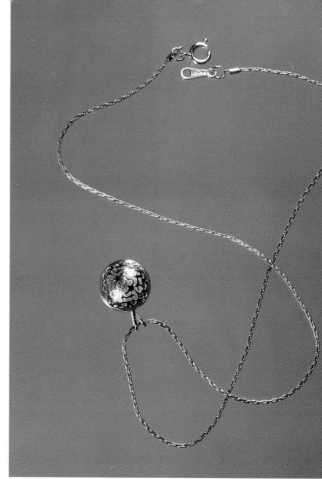

Aqua filigree pendant with chain, 16-inch chain, 1984. *Courtesy of Mary Jo Michealis.* $25-35.

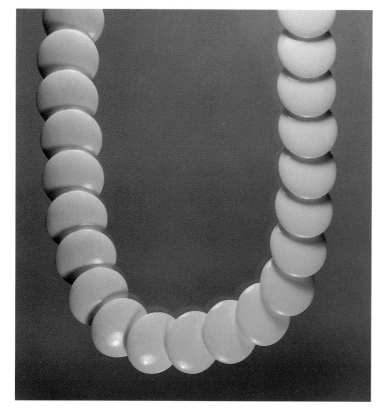

White plastic necklace, circa 1980s. *Courtesy of Sarah Jones.* $20-30.

Gold 24-inch necklace with letter "M" and pearl, circa 1980s. *Courtesy of Mary Jo Michealis.* $20-30.

Birthstone pendant with ruby stone, 18.5-inch chain, circa 1980s. *Courtesy of Mary Jo Michealis.* $20-30.

Hand-painted porcelain locket depicting floral motif surrounded by seed pearls, 24-inch chain, circa 1980s. *Courtesy of Mary Jo Michealis.* $45-55.

Plastic 24-inch necklace, circa 1970s. *Courtesy of Kenneth L. Surratt, Jr.* $20-30.

Reversible Cameo Pendant Necklace, 30-inch chain, circa 1974. *Courtesy of Mary Jo Michealis.* $45-55.

Gold necklace with three hearts to be worn together or separately, 28-inch chain, circa 1990s. *Courtesy of Kim Gibson.* $30-40.

Gold butterfly 30-inch necklace, circa 1980s. *Courtesy of Mary Jo Michealis.* $20-30.

Gold 19-inch butterfly necklace, circa 1980s. *Courtesy of Mary Jo Michealis.* $15-20.

Perfume bottle pendant, 32-inch chain, circa 1980s. *Courtesy of Mary G. Moon.* $35-45.

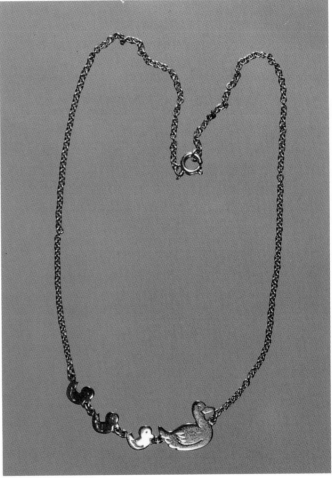

Gold necklace with duck and three ducklings, 13 inches, circa 1980s. *Courtesy of Mary Jo Michealis.* $25-35.

Back of perfume bottle pendant necklace.

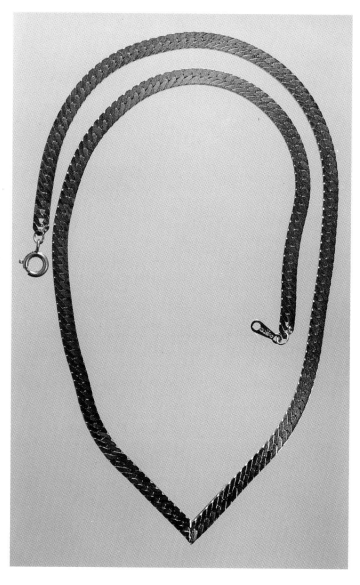

Silver 20-inch necklace, circa 1980s. *Courtesy of Mary G. Moon.* $20-25.

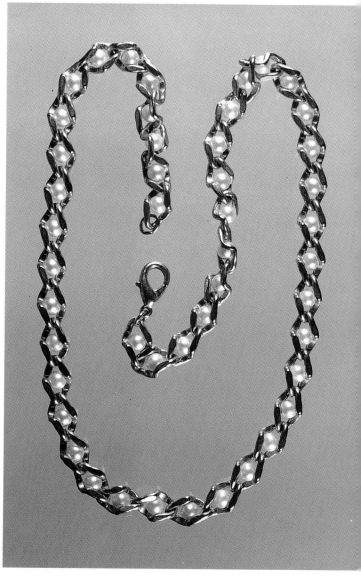

Pearl and gold necklace, 18 inches, circa 1990s. *Courtesy of Kim Gibson.* $35-45.

Etched Austrian Lead crystal pendant depicting the Great Oak, Tree of Life motif, silver 24-inch chain, 1981. *Courtesy of Miss Mary Jane.* $45-55.

Left: Gold necklace, 31-inch chain, circa 1990s. $20-25.
Center: Gold necklace with initial "L", 24-inch chain, circa 1990s. $18-22.
Right: Gold pendant depicting Demeter (Ceres), goddess of the harvest with sheath of wheat, 19-inch chain, circa 1990s. $30-40.
Necklaces courtesy of Mary G. Moon.

Heart pendant with pearls and amethyst stone, 17-inch chain, circa 1970s. *Courtesy of Mary G. Moon.* $25-35.

Pink plastic heart pendant, 16-inch chain, circa 1980s. *Courtesy of Miss Mary Jane.* $15-25.

Left: Gold heart with pearls and ruby, 18-inch chain, circa 1990s. $25-35.
Center: Gold pendant with clear stones, circa 1990s. $30-40.
Right: Pendant with square ruby red stone, 18-inch chain, circa 1980s. $30-40.
Necklaces courtesy of Miss Mary Jane.

Pendant with garnet stone, 17-inch chain, circa 1990s. *Courtesy of Miss Mary Jane.* $45-55.

Gold cross with rhinestones, 18-inch chain, circa 1990s. *Courtesy of Miss Mary Jane.* $35-45.

Pendant with amethyst stone, 19-inch chain, circa 1990s. *Courtesy of Miss Mary Jane.* $45-55.

Silver pendant with rose, 24-inch chain, circa 1990s. *Courtesy of Miss Mary Jane.* $35-45.

Left: Gold heart pendant with ruby red stones, 18 inches, circa 1990s. $25-35.
Center: Avon Romantic Birthstone Pendant with loose ruby red stone, 24 inches, circa 1990s. $20-25.
Right: Poinsettia Heart Pendant with oval ruby red stones and rhinestones, 18 inches, circa 1990s. $25-35.
Pendant necklaces courtesy of Miss Mary Jane.

Kenneth Jay Lane Poetic Romance Cameo Pin/Pendant Necklace with pearls and rhinestones, 22-inch chain with pearls, 1994. *Courtesy of Betty Barry.* $65-75.

Imari Simulated Pearls in soft black, white, and sweet cream with Imari enhancer, each strand measures 36 inches, 1985. *Courtesy of Sarah Jones.* $45-55.

Chapter Nine
Related Awards and Gifts

Mrs. Albee Award depicting woman wearing blue and yellow, 1980. *Courtesy of Miss Mary Jane*. $175-195.

Mrs. Albee Award depicting woman in brown, 1981. *Courtesy of Miss Mary Jane*. $125-175.

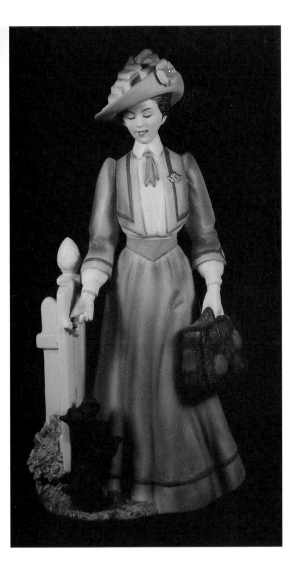

Mrs. Albee Award depicting woman wearing lavender dress, 1983. *Courtesy of Miss Mary Jane.* $95-125.

Mrs. Albee Award depicting Mrs. Albee wearing pink and gray, circa 1984. *Courtesy of Miss Mary Jane.* $125-150.

Mrs. Albee Award depicting Mrs. Albee seated at desk, 1985. *Courtesy of Miss Mary Jane.* $150-175.

Another view of Mrs. Albee Award, 1985.

Mrs. Albee Centennial Award depicting Mrs. Albee in formal attire, circa 1986. *Courtesy of Miss Mary Jane.* $125-195.

Mrs. Albee Award depicting Mrs. Albee with umbrella standing next to mailbox, 1988. *Courtesy of Miss Mary Jane.* $175-195.

Mrs. Albee Award depicting Mrs. Albee wearing green and black dress holding umbrella, 1987. *Courtesy of Miss Mary Jane.* $85-95 as is.

Another view of Mrs. Albee Award, 1987.

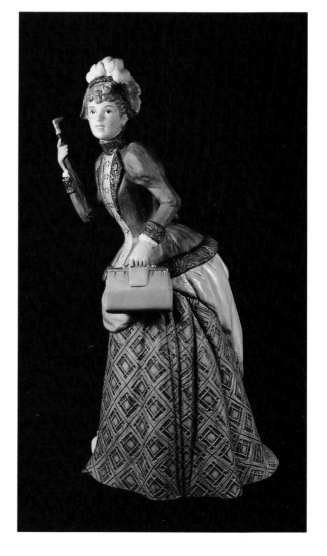

Mrs. Albee Award depicting Mrs. Albee in red with basket and packages, 1989. *Courtesy of Miss Mary Jane.* $175-195.

Mrs. Albee Award depicting Mrs. Albee in yellow dress, 1990. *Courtesy of Miss Mary Jane.* $125-175.

Mrs. Albee Award depicting Mrs. Albee with autumn motif, 1991. *Courtesy of Miss Mary Jane.* $195-225.

Another view of Mrs. Albee Award, 1991.

Mrs. Albee Award depicting Mrs. Albee seated in blue dress, 1992. *Courtesy of Miss Mary Jane.* $125-175.

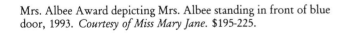

Mrs. Albee Award depicting Mrs. Albee standing in front of blue door, 1993. *Courtesy of Miss Mary Jane.* $195-225.

Mrs. Albee Award depicting Mrs. Albee wearing red dress and holding fan, 1994. *Courtesy of Miss Mary Jane.* $175-195.

Mrs. Albee Award depicting Mrs. Albee with teacup, circa 1995. *Courtesy of Miss Mary Jane.* $175-195.

Mrs. Albee Award depicting Mrs. Albee wearing blue and lavender dress, 1996. *Courtesy of Miss Mary Jane.* $175-195.

Mrs. Albee Award depicting Mrs. Albee wearing red dress and poodle with package, 1997. *Courtesy of Miss Mary Jane.* $175-195.

1985–1986 Silver Mrs. Albee Award.
Courtesy of Miss Mary Jane. $175-195.

1984–1985 Gold Mrs. Albee Award.
Courtesy of Miss Mary Jane. $195-225.

Silver Mrs. Albee District Award, 1987.
Courtesy of Miss Mary Jane. $150-200.

Left: Miniature Mrs. Albee Award gift, 1994. $25-35.
Center: Miniature Mrs. Albee Award gift, 1993. $25-35.
Right: Miniature Mrs. Albee Award gift, 1990. $25-35.
Miniature Mrs. Albee Award gifts courtesy of Miss Mary Jane.

Left: Miniature Mrs. Albee Award gift, 1986. $25-35.
Middle. Miniature Mrs. Albee Award gift, 1989. $25-35.
Right: Miniature Mrs. Albee Award gift, 1988. $25-35.
Miniature Mrs. Albee Award gifts courtesy of Miss Mary Jane.

Left: Miniature Mrs. Albee Award gift on green pedestal, 1996. $25-35.
Right: Miniature Mrs. Albee Award gift on blue pedestal, 1995. $25-35.
Miniature Mrs. Albee Award gifts courtesy of Miss Mary Jane.

President's Club Award Lenox bowl, 1980. *Courtesy of Miss Mary Jane.* $50-60.

Bottom of President's Club Award Lenox bowl.

Second Avon Anniversary Award porcelain plate with doorknocker, 1979. *Courtesy of Miss Mary Jane.* $20-30.

Wedgwood Fifth Avon Anniversary Plate with Great Oak, 1978. *Courtesy of Miss Mary Jane.* $35-45.

15th Anniversary Avon Plate Award, 1989. *Courtesy of Miss Mary Jane.* $20-30.

Tenth Avon Anniversary Plate with California Perfume Company design, 1979. *Courtesy of Miss Mary Jane.* $25-35.

20th Anniversary Avon Plate Award with likeness of Mrs. Albee, 1993. *Courtesy of Miss Mary Jane.* $20-30.

Award Clock, 1990s. *Courtesy of Betty Barry.* $25-35.

Back of clock.

My First Call Precious Moments Award, 1980. *Courtesy of Miss Mary Jane.* $45-55.

The Day I Made President's Club Precious Moments Award, President's Club Luncheon 1980. *Courtesy of Miss Mary Jane.* $35-45.

Come Rain or Shine Precious Moments Award, 1993. *Courtesy of Miss Mary Jane.* $40-50.

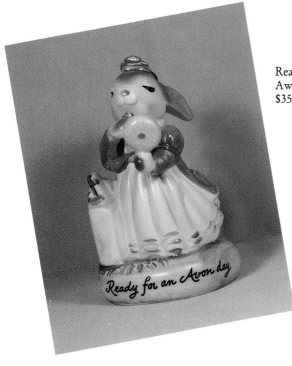

Ready for an Avon Day Precious Moments Award, 1980. *Courtesy of Miss Mary Jane.* $35-45.

Side view of Ready for an Avon Day Award.

Which Shade Do You Prefer? Precious Moments Award, 1980. *Courtesy of Miss Mary Jane.* $65-75.

Honor Society Cup and Saucer Award with pink roses and likeness of Mrs. Albee, 1991. *Courtesy of Miss Mary Jane.* $40-50.

Honor Society Cup and Saucer Award with pale pink roses and likeness of Mrs. Albee, 1992. *Courtesy of Miss Mary Jane.* $40-50.

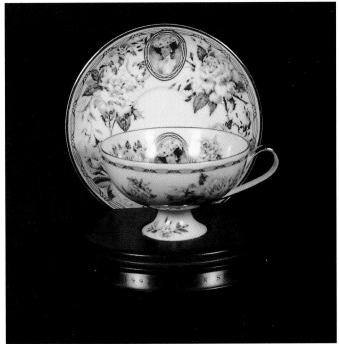

Honor Society Cup and Saucer Award with small likeness of Mrs. Albee, 1993. *Courtesy of Miss Mary Jane.* $35-45.

Honor Society Cup and Saucer with red border, 1994. *Courtesy of Miss Mary Jane.* $40-50.

Honor Society Cup and Saucer Award with blue and gold border, 1995. *Courtesy of Miss Mary Jane.* $45-55.

Honor Society Cup and Saucer Award with white background and floral bouquet, 1996. *Courtesy of Miss Mary Jane.* $45-55.

Endnotes

[1] Hastin, Bud. *Bud Hastin's Avon & C.P.C. Collector's Encyclopedia* (Ft. Lauderdale, Florida: Bud Hastin, 1995), 10-11.

[2] Ibid., p. 11.

[3] Ibid.

[4] Schneider, Dee. *President's Club Jewelry* (Glendale, California: Avons Research Publications, 1974), 14.

[5] Ibid.

[6] Hastin, *Bud Hastin's Avon & C.P.C. Collector's Encyclopedia*, 15.

[7] Schneider, Dee. *President's Club Jewelry* (Glendale, California: Avons Research Publications, 1974), p. 3.

[8] Hastin, *Bud Hastin's Avon & C.P.C. Collector's Encyclopedia*, p. 15

[9] Ibid.

[10] Ibid.

[11] Morris, Betsy, "If Women Ran the World It Would Look a Lot Like Avon," *Fortune*, 21 July 1997, 76.

[12] Ibid., 78.

[13] Ibid.

[14] Ibid., 76.

[15] Ibid.

[16] Ibid., 74.

[17] Ibid.

[18] Ibid., 76.

[19] *Hagley Museum and Library Newsletter*, Summer 1997.

[20] Schneider, Dee. *President's Club Jewelry*, p. 7.

[21] Schneider, Dee. *President's Club Jewelry*, p. 9.

[22] Schneider, Dee. *President's Club Jewelry*, p. 7.

[23] Schneider, Dee. *President's Club Jewelry*, p. 23.

[24] Ibid.

[25] Schneider, Dee. *President's Club Jewelry*, p. 23.

[26] Schneider, Dee. *President's Club Jewelry*, p. 25.

[27] *Members Only: Avon's Magazine for President's Club*, Spring 1985.

Bibliography

Avon Products, Inc. special advertising section, "The Elizabeth Taylor Fashion Jewelry Collection," *New Woman*, December 1993.

Ettinger, Roseann. *Popular Jewelry of the '60s, '70s, &'80s*. Atglen, Pennsylvania: Schiffer Publishing Ltd., 1997.

Hagley Museum and Library Newsletter, Summer 1997.

Hastin, Bud. *Bud Hastin's Avon & C.P.C. Collector's Encyclopedia*. Ft. Lauderdale, Florida: Bud Hastin, 1995.

Members Only: Avon's Magazine for President's Club, Spring 1985.

Morris, Betsy, "If Women Ran the World It Would Look A Lot Like Avon," *Fortune*, July 21, 1997.

Schiffer, Nancy. *The Best of Costume Jewelry*. West Chester, Pennsylvania: Schiffer Publishing Ltd., 1990.

Schneider, Dee. *Avon's Award Bottles, Gifts, Prizes*. Glendale, California: Avons Research Publications, 1975.

Schneider, Dee. *President's Club Jewelry*. Glendale, California: Avons Research Publications, 1974.